STOCHASTIC GAMES AND RELATED TOPICS

THEORY AND DECISION LIBRARY

General Editors: W. Leinfellner and G. Eberlein

Series A: Philosophy and Methodology of the Social Sciences
Editors: W. Leinfellner (Technical Universtiy of Vienna)
G. Eberlein (Technical University of Munich)

Series B: Mathematical and Statistical Methods
Editor: H. Skala (University of Paderborn)

Series C: Game Theory, Mathematical Programming and Operations Research
Editor: S. H. Tijs (University of Nijmegen)

Series D: System Theory, Knowledge Engineering and Problem Solving
Editor: W. Janko (University of Vienna)

SERIES C: **GAME THEORY, MATHEMATICAL PROGRAMMING AND OPERATIONS RESEARCH**

Editor: S. H. Tijs (University of Nijmegen)

Volume 7

Scope

Particular attention is paid in this series to game theory and operations research, their formal aspects and their applications to economic, political and social sciences as well as to socio-biology. It will encourage high standards in the application of game theoretical methods to individual and social decision making.

The titles published in this series are listed at the end of this volume.

STOCHASTIC GAMES AND RELATED TOPICS

In Honor of Professor L. S. Shapley

Edited by

T. E. S. RAGHAVAN

University of Illinois, Chicago, U.S.A.

T. S. FERGUSON

University of California, Los Angeles, U.S.A.

T. PARTHASARATHY

Indian Statistical Institute, India

and

O. J. VRIEZE

University of Limburg, The Netherlands

SPRINGER SCIENCE+BUSINESS MEDIA, B.V.

Library of Congress Cataloging-in-Publication Data

Stochastic games and related topics : in honor of Professor L.S.
Shapley / edited by T.E.S. Raghavan ... [et al.].
p. cm. -- (Theory and decision library. Series C, Game
theory, mathematical programming, and operations research ; v. 7)
Includes bibliographical references.
ISBN 978-94-010-5673-1 ISBN 978-94-011-3760-7 (eBook)
DOI 10.1007/978-94-011-3760-7
1. Game theory. 2. Stochastic processes. I. Shapley, Lloyd S.,
1923- . II. Raghavan, T. E. S. III. Series.
QA269.S76 1990
519.3--dc20 90-48537

ISBN 978-94-010-5673-1

Printed on acid-free paper

CONTENTS

Acknowledgement by the Editors vii

Acknowledgement by *T.E.S. Raghavan* viii

Biographical Note on Lloyd S. Shapley by *T.S. Ferguson* ix

Stochastic games - an overview by *T.E.S. Raghavan* 1

A brief summary of the papers in the volume by *T.E.S. Raghavan* 11

Contributed Papers

1 *Christopher P. Ferguson and Thomas S. Ferguson*, Models for the Game of Liar's dice. U.C.L.A. 15

2 *William H. Ruckle*, A Discrete Search Game. Clemson University. 29

3 *Michèle Breton*, Algorithms for Stochastic Games. GERAD Ecole des Hautes Etudes Commerciales de Montreal, Canada. 45

4 *Jerzy A. Filar and Boleslaw Tolwinski*, On the Algorithm of Pollatschek and Avi-Itzhak. University of Maryland at Baltimore and Colorado School of Mines. 59

5 *Sagnik Sinha, Frank Thuijsman and Stef H. Tijs*, Semi-infinite Stochastic Games. Indian Statistical Institute, University of Limburg, The Netherlands and Catholic University of Nijmegen, The Netherlands. 71

6 *Frank Thuijsman and Okko J. Vrieze*, Easy Initial States in Stochastic Games. University of Limburg, The Netherlands. 85

7 *Sylvain Sorin and Shmuel Zamir*, Big Match with Lack of Information on One Side (III). University of Strassbourg, France and University of Pittsburg. 101

8 *Costis Melolidakis*, Stochastic Games with Lack of Information on One Side and Positive Stop Probabilities. Technical University of Crete. 113

9 *Andrzej S. Nowak and T. E. S. Raghavan*, Positive Stochastic Games and a Theorem of Ornstein. Wroclaw Technical University and the University of Illinois at Chicago. 127

10 *Andrzej S. Nowak*, Existence of Correlated Weak Equilibria in Discounted Stochastic Games with General State Space. Wroclaw Technical University, Poland. 135

11 *Jean-Francois Mertens and T. Parthasarathy*, Nonzero-Sum Stochastic Games. C.O.R.E., Belgium and Indian Statistical Institute, India. 145

12 *Rabah Amir,* On Stochastic Games with Uncountable State and Action Spaces. S.U.N.Y. at Stony Brook. 149

13 *Ulrich Rieder,* Non-Cooperative Dynamic Games with General Utility Functions. University of Ulm, West Germany. 161

14 *Mukul Majumdar and Rangarajan Sundaram,* Symmetric Stochastic Games of Resource Extraction: The Existence of Non-Randomized Stationary Equilibrium. Cornell University, Ithaca. 175

15 *Xiong Zhong Kai, George Leitmann and Franco Garofalo,* Robustness of Uncertain Dynamical Systems with Delay in the Absence of Matching Assumptions. Chongquing Communication Institute, People's Republic of China, University of California at Berkeley, Universita di Napoli, Italy. 191

Ancestral papers

Lloyd S. Shapley (1953). Stochastic Games, Proc. Nat. Acad. Sciences, U.S.A., **39**, 1095-1100. 201

John Milnor and Lloyd S. Shapley (1957). On Games of Survival, Ann. Math. Studies, **39**, 15-45, Princeton University Press. 207

ACKNOWLEDGEMENT I

The origins of this volume can be traced back to the workshop on "Stochastic Games " that was held at the University of Illinois at Chicago during June 26- 28, 1987. Sixteen researchers participated in the workshop presenting their most recent contributions in this growing field of importance within Game Theory. Originally the plan was to just bring out a conference volume reflecting the presentations at the workshop. When the participants finally decided that such a volume should be dedicated to Professor L.S.Shapley, it was agreed upon by one and all that the volume should contain only original and unpublished articles of their most significant research in the area. Needless to say, that all the papers submitted went through the normal refereeing process, usually by two referees, before they were accepted for the volume.

The editors owe a lot to many distinguished researchers in the field who shared the responsibility of refereeing these papers for the volume. We felt it appropriate to include the seminal paper by Professor Shapley on "Stochastic Games ", Proc. National Academy of Sciences, 1953, that triggered all these developments. We also felt that it would be very useful and appropriate to include here the closely related paper by Professors Milnor and Shapley on "Games of Survival" that focuses on many other complex issues in models of multistage games, another rich source for further developments in multistage games. We wish to acknowledge the kind permission from Professors Milnor and Shapley and also the permission from the Princeton University Press for allowing the inclusion of the above two papers in this volume.

We thank the authors for joining the editors in honoring our distinguished Game Theorist Professor Lloyd S. Shapley by publishing their best recent researches in the area in the honor volume.

We wish to acknowledge the support of the University of Illinois at Chicago and the Institute of Applied Mathematics at Minnesota, for providing funds to organize the International Workshop on Stochastic Games.

Our special thanks to Ms Eva Roper, Ms Mahalia Triplett and Mrs Marlies Haenen for their excellent retyping of the accepted papers into a uniform style.

T.S.F.
T.P.
T.E.S.R.
O.J.V.

ACKNOWLEDGEMENT II

While coordinating the editorial work, here at Chicago, I had to continuously seek the help of many professional colleagues, departmental staff members and friends and even old classmates at various stages of this volume. This list is too numerous to include here. Yet my sense of gratitude dictates that I must acknowledge Stef Tijs, John Wood, Andrzej Nowak, Swaminathan Sankaran, Marty Gartzman, Mario Widel, John Cepek, Aziz Kotob, Sam Eid, Chih-Ru Hsio, Frank Thuijsman, and Jean Derks for their enthusiastic help at various stages.

The commitment to bring out the typed version to uniform style meant more than doing the uniform TEX typing. I needed someone who would spend more or less full time working on the font changes, massive proof reading and looking for missing math symbols etc. Only my heart can tell what it means to discover persons like Tajudeen Sokoya and Evangelista Fe who cheerfully took on their own shoulders a lot of my responsibilities.

T.E.S.Raghavan
August 9, 1989

BIOGRAPHICAL NOTE ON

Lloyd S. Shapley

It has been over 35 years since a modest paper entitled "Stochastic Games" appeared in the Proceedings of the National Academy of Sciences. Yet even today, that paper is remarkable for its originality and its generality. It defines finite stochastic games and proves the existence of a value and stationary optimal strategies in the case where all stop probabilities are positive. The method of proof yields an iterative technique for approximating the solution that is still competitive today. It appeared at a time when not even the one person version of the problems, known today as Markov Decision Processes, had been developed. It also contains the solution of ratio games as a special case. Its impact will be obvious to anyone reading this volume, devoted to various aspects of the area of mathematics that originated with the publication of that paper. This volume is intended to honor the author of "Stochastic Games", Lloyd S. Shapley.

Lloyd Shapley was born on June 2, 1923, in Cambridge, Massachusetts, one of the sons of the distinguished astronomer Harlow Shapley. He was a student at Harvard when he was drafted in 1943, and in the same year, as a sergeant in the Army Air Corps in Sheng-du, China, he received the Bronze Star decoration for breaking the Japanese weather code. In 1945, he returned to Harvard, where he graduated in 1948 with an A.B. in Mathematics. After working briefly at the Rand Corporation in Santa Monica, he returned to the East Coast in 1949 to attend graduate school at Princeton University, where he obtained a Ph.D. in Mathematics in 1953 under the direction of A. W. Tucker. Among the graduate students attending Princeton at that time and sharing Shapley's interest in game theory were D. Gale, H. W. Kuhn, J. W. Milnor, and J. F. Nash. In 1954, he joined the active game theory group at the Rand Corporation which included D. Blackwell, H. F. Bohnenblust, M. Dresher, S. Karlin. J. C. C. McKinsey, and J. D. Williams. In 1955, he became a research fellow at the California Institute of Technology where he worked for one year before returning to the Rand Corporation. In the course of the same year, Marion Ludolph, a mathematician and coworker at Rand, became his wife. They have two children, Peter and Christopher. As his outstanding contributions brought him increasing recognition, he was elected a Fellow of the Econometric Society in 1967 and of the American Academy of Arts and Science (founded by his father) in 1974, and in 1978 he was elected to the National Academy of Science. In 1981 he received a joint appointment in the Department of Economics and Mathematics at the University of California in Los Angeles and, in the same year, he was awarded the John von Neumann Prize in Operations Research and Management Science. In 1986, he received an Honorary Doctorate from the Hebrew University in Jerusalem.

Professor Shapley is one of those rare individuals of whom it may be said that an entire field, in this case game theory, would be different without him. He has been directly involved in most of the important developments in this subject since 1950, was instrumental in many of them, and is as active as ever. In contrast to his introduction of the topic of stochastic games whose study continues unabated today, his early work with Snow and Karlin on matrix games was so complete that little has been added since. He has also made fundamental contributions to cooperative game theory. His value concept for n person games, now known as the Shapley value, is the most extensively studied solution concept in game theory. He has been instrumental in the development of utility theory, and it was he who laid much of the groundwork for the solution of the problem of the existence of Von Neumann-Morgenstern stable sets. His work with Gale on assignment problems such as marriage or college admissions, his work on balanced collections of coalitions with applications to characterizing the core of an arbitrary game and to characterizing market games, his work with Aumann on nonatomic games, his work with Maschler and Peleg on the kernel and the nucleolus, have all had a tremendous impact in economic theory. Further, he has been at the forefront of developing and applying game theoretical tools to economics ("equivalence theorems", for example) and political science (the "power index").

T.S. Ferguson
Los Angeles
March 1989

Stochastic Games - An Overview

T.E.S. Raghavan

1. Finite State Stochastic Games : A zero-sum stochastic game as formulated by Shapley [39] is played in stages. At each stage, the game is in one of finitely many states, $s = 1, 2, \ldots, S$, in which players I and II are obliged to play a matrix game, $A^s = (a^s_{ij})$ $s = 1, \ldots, S$, once. The Markovian "law of motion" is defined by transition probabilities, $q(s'/s, i, j)$, where $\{s'/s, i, j\}$ is the event that the game will enter state s' at the next stage given that at the current stage the state of the game is s, and I and II choose the ith row and the jth column of A, respectively. In general, players' strategies may depend on complete past histories. In particular, a strategy f for player I is a sequence $(f_1, f_2, .., f_n, ..)$ where f_n associates some probability distribution on the rows of A^{s_n} with each n-step past history $h_n = (s_1, i_1, j_1, s_2, i_2, j_2, \ldots, s_n)$. Here (i_k, j_k) denotes players' choices in s_k. A strategy $f = (f_1, f_2, \ldots)$ is called *stationary* if $f_1 = f_2 = \ldots$ where the common f_i's are conditioned only on the current state of the game. For a given strategy pair (f, g), let $r_n(f, g)(s)$ denote the expected income to player I at the nth stage when the game begins in state s. The overall payoff for the players in the stochastic game is determined by the manner in which the players evaluate a stream of immediate rewards $(r_1, r_2, \ldots)$. In a *positive stop probability* stochastic game, as originally defined by Shapley, player I continues to accumulate the immediate rewards. However, from any state the game could enter a terminal state with a state and actions dependent probability $1 - \beta > 0$. Once the terminal state is reached, the game is over and no more accumulations are possible. If β is independent of the states and actions an alternative formulation of this game will be to discount future payoffs at a fixed discount rate β. The payoffs in such β discounted stochastic games are defined by

$$\phi_\beta(f,g)(s) = \sum_{n=1}^{\infty} \beta^{n-1} r_n(f,g)(s),$$

where $\phi_\beta(f,g)(s)$ is the aggregated expected income to player I when s is the initial state of the game and players I and II use the strategy pair (f,g).

An *undiscounted payoff* is a functional on the strategy spaces, given by

$$\phi(f,g)(s) = \liminf_{N\to\infty} \frac{1}{N} \sum_{n=1}^{N} r_n(f,g)(s),$$

where $\phi(f,g)(s)$ measures the long run expected reward per play. In case the immediate rewards are nonnegative, one can consider just the total payoff over the infinite play. This is well defined, but could be infinite. Games with total payoffs and nonnegative immediate rewards are called *positive stochastic games*. One could choose many other functionals aggregating an infinite stream of payoffs such as the limit superior of the long run average, or a convex combination of the discounted and the undiscounted payoffs etc.

A nonzero-sum stochastic game with two players consists of separate immediate rewards a_{ij}^s, b_{ij}^s, one for each player, with a law of motion $q(t/s,i,j)$ among states. For any strategy pair (f,g) we can separately evaluate the expected immediate rewards at the nth stage. As in zero-sum stochastic games, the β-discounted and the undiscounted payoffs, can also be defined for each player. We will denote by $\phi_\beta^i(f,g)$ the β-discounted reward and $\phi^i(f,g)$ the undiscounted reward to player i, where $i =$ I or II.

A *β-discounted* zero-sum stochastic game has the value $v_\beta(s)$ if and only if

$$\inf_g \sup_f \phi_\beta(f,g)(s) = \sup_f \inf_g \phi_\beta(f,g)(s) = v_\beta(s), \quad s = 1,2,..,S.$$

The value $v(s)$ for an undiscounted zero-sum stochastic game is similarly defined.

For a β-discounted nonzero-sum stochastic game, a pair of strategies (f^*,g^*) constitutes a Nash equilibrium point if and only if

$$\phi_\beta^I(f^*,g^*)(s) \geq \phi_\beta^I(f,g^*)(s) \text{ for all } f \text{ at each } s = 1,2,\ldots,S,$$
$$\phi_\beta^{II}(f^*,g^*)(s) \geq \phi_\beta^{II}(f^*,g)(s) \text{ for all } g \text{ at each } s = 1,2,\ldots,S.$$

The definition also extends to undiscounted payoffs $\phi^I(f,g)(s)$ and $\phi^{II}(f,g)(s)$.

As alternative solution concepts, one could also consider other equilibrium notions such as the concept of a correlated equilibrium point due to Aumann [2] or the notion of a weak equilibrium point due to Moulin and Vial [25].

2. Structured Stochastic Games:

We can classify stochastic games based on additional properties of the action spaces, immediate rewards, and transition probabilities. If the game has *perfect information*, then one of the players will have a singleton as his action space. The game is called *irreducible* if the transition matrix is an irreducible Markov chain for any arbitrary pair of stationary strategies for the two players. The law of motion is said to be *additive* if and only if $q(t/s,i,j) = q_1(t/s,i) + q_2(t/s,j)$ for all s,i,j for some q_1, q_2. We call a stochastic game a *switching control* stochastic game if the law of motion is additive and at each state s either $q_1 = 0$ or $q_2 = 0$. The game is called a *single controller game* if and only if the transition is additive and, further, $q \equiv q_1$ or $q \equiv q_2$ for all s. Thus, when the game is controlled by say player II, then $q(t/s,i,j) = q_2(t/s,j)$ for all s,t,i,j. The law of motion is said to be *state independent* if and only if $q(t/s,i,j) \equiv q(t/i,j)$ for all s,t. The rewards for player I are called *separable* if and only if the immediate rewards $a^s_{ij} = r(s,i,j)$ satisfy $r(s,i,j) = r_1(s) + r_2(i,j)$ for all s,i,j. The rewards are called additive if $r(s,i,j) = r_1(s,i) + r_2(s,j)$.

3. Orderfield Property:

Let $\mathcal{F}$ be the smallest Archimedean ordered field containing the data of a mathematical problem with finitely many entries. We say that the problem possesses *Orderfield Property* if and only if the components of some solution to the problem also lie in $\mathcal{F}$.

4. Some Historical Remarks :

Stochastic games were first formulated by Shapley in 1953. In his seminal paper, Shapley [39] established the existence of value and optimal stationary strategies for zero-sum β-discounted games. His proof actually gave a recipe for the iterative computation of the value. Gillette [13] introduced zero-sum undiscounted stochastic games and pointed out that in general they may not possess optimal stationary strategies. Blackwell and Ferguson [5] studied Gillette's example and showed that this game does possess an ϵ-optimal *behavior strategy* for Player I and a stationary optimal strategy for player II. Kohlberg [18] extended their specific solution to stochastic games with exactly one nonabsorbing state. Monash, in his Ph.D. thesis, [24] attempted to prove the existence of value in behavior strategies for all undiscounted stochastic games. Using the Puiseux series technique developed by Bewley and Kohlberg [3], [4], Mertens and Neyman proved [22] that all undiscounted zero-sum stochastic games with finitely many states and finitely many actions have value in behavior strategies.

The existence of Nash equilibria in stationary strategies for discounted nonzero-sum stochastic games were independently established by Fink [10], Takahashi [43], Rogers [38], and Sobel [40]. Undiscounted nonzero-sum games were first studied by Federgruen [7]. Parthasarathy and Raghavan [32] showed that single controller stochastic games have undiscounted stationary equilibria. Recently, Vrieze and Thuijsman [48] have proved that undiscounted nonzero-sum stochastic games with exactly one non absorbing state admit ϵ-equilibria in behavior strategies.

5. Algorithmic Aspects:

Naturally one would look for games where the optimal strategies are simpler. Besides discounted stochastic games, many classes of structured stochastic games in undiscounted payoffs such as games with perfect information [20], games with irreducible transitions, [16], one player control games ([42], [32]), switching control games [8], additive reward and additive transition games [35]), and separable reward-state independent transition games [33], do possess stationary optimals. Among them many are also known to possess order field property [36]. Recently, Filar et al. [9] have given a characterization of undiscounted games admitting stationary optimals via certain nonlinear programs.

Attempts to develop efficient finite step algorithms for special classes have resulted in a new line of research. One player control discounted as well as undiscounted zero-sum games are reducible to a single linear program ([32], [17], [46]). In addition, there are finite step algorithms to solve many of the structured stochastic games ([9], [23], [33], [35], [47], [48]). Little is known about nonzero-sum stochastic games with undiscounted payoffs, except for the unichained games [7], and one player control games [32]. These two classes admit stationary equilibria. Further, one could also compute a Nash equilibrium pair, in finitely many steps, for one player control games [29]. The details of many of these algorithms are surveyed in [36].

Many iterative algorithms for general zero-sum stochastic games have been proposed in the literature. Hoffman and Karp [16] proposed the first such algorithm for undiscounted irreducible stochastic games. Pollatschek and Avi-Itzhak [34] proposed another algorithm along the lines of the Newton-Raphson procedure. For some remarks on this algorithm, see [44]

6. Stochastic Games in Arbitrary State Spaces:

A stochastic game in an arbitrary state space and compact action spaces consists of the following objects: **1.** State space S- an arbitrary Borel subset of a complete separable metric space. **2.** A finite set $N = 1, 2, \ldots, n$ of players. **3.** Compact action spaces $A_1, A_2, \ldots, A_n$, one for each player. **4.** Immediate reward functions $r_i(s, a_1, a_2, \ldots, a_n)$, $i = 1, 2, \ldots, n$ where the r_i's are bounded measurable functions on $S \times A_1 \times \ldots \times A_n$. **5.** Markovian transition $q(dt/s, a_1, a_2, \ldots, a_n)$ where q is a regular conditional probability measure, namely for each $u = (s, a_1, \ldots, a_n)$, $q(dt, u)$ is a probability measure on S and $q(E/u)$ is a measurable function of u for each fixed Borel set $E \subset S$. **6.** For discounted games, a discount factor $0 \leq \beta < 1$.

The actual realization of the game is similar to finite state stochastic games in discrete time. On the kth day, the players could use all the information so far available to them in choosing actions on the partial history $h_k =$

$(s_1, a_1^1, \ldots, a_n^1; s_2, a_1^2, \ldots, \ldots, a_n^2; \ldots, s_k)$. A strategy $f^i = \{f_k^i\}$, for player i, $i = 1, 2, \ldots, n$, can select a probability measures on A_i, one for each history h_k, in such a way that f_k^i is a measurable function of h_k.

With these preliminaries we can formally define the notions of Nash equilibria or correlated equilibria. In an obvious fashion as in general N-person games, ϵ-Nash equilibria and ϵ-correlated equilibria can also be defined for each starting state.

Zero-sum discounted stochastic games in arbitrary state spaces were first studied by Maitra and Parthasarathy [21]. Under some regularity conditions, they proved the existence of value in stationary strategies. Positive stochastic games were studied by many authors ([12], [14], [19], [26]). They proved the existence of value and an optimal stationary measurable strategy for the minimizer for all $\beta \leq 1$.

7. Nonzero-Sum Discounted Stochastic Games in Arbitrary State Spaces:

Nonzero-sum discounted stochastic games in arbitrary state spaces were first studied by Himmelberg, Parthasarathy, Raghavan, and Van Vleck [14]. With additive rewards and additive transitions and with some regularity conditions on the transitions they showed the existence of stationary p-equilibria, which was shown to be equivalent to establishing ordinary stationary Nash equilibria [31]. Under very general regularity conditions, Rieder [37], Whitt [49], and Nowak [27] have proved the existence of ϵ-Nash stationary equilibria for discounted stochastic games in arbitrary state spaces.

As an alternative, one could consider other equilibrium concepts that are equally attractive and relevant to the problems of nonzero-sum games. Under mild regularity conditions Nowak and Raghavan [28] have shown that in an arbitrary state space, with compact action spaces, and continuous density transitions the players have a *correlated equilibrium in stationary strategies* which is also a *communication equilibrium* [11].

References

[1] *Aumann, R.J.* (1981). *Survey of Repeated Games,* "Essays in Game Theory and Mathematical Economics." Bibliographisches Institute, Mannheim, W. Germany.

[2] *Aumann, R.J.* (1974). Subjectivity and Correlation in Randomized Strategies. *J. Math. Economics,* **1**, 67-96

[3] *Bewley, T. and E. Kohlberg* (1976). The Asymptotic Theory of Stochastic Games. *Math. Oper. Res.,* **1**, 197-208, 321-336.

[4] *Bewley, T. and E. Kohlberg* (1978). On Stochastic Games with Stationary Optimal Strategies, *Math. of Oper. Res.,* **2**, 104-125.

[5] *Blackwell, D. and T. S. Ferguson* (1968). The Big Match, *Ann. Math. Stat.*, **39**, 159-163.

[6] *Everett, H.* (1957). Recursive Games, "Contributions to the theory of games", Vol III: Edited by M. Dresher, A.W. Tucker and P. Wolf, Ann. Math Studies, **39**, 47-78, Princeton University Press, Princeton, New Jersey.

[7] *Federgruen, A.* (1978). Markovian Control Problems, Ph.D. Thesis, Mathematisch Centrum, Amsterdam.

[8] *Filar, J.A.* (1981). Order Field Property for Stochastic Games when the Player who Controls Transitions Changes from State to State, *J. Optim. Theo. App.*, **34**, 505-517.

[9] *Filar, J.A., T.A. Schultz, F. Thuijsman and O.J. Vrieze* (1987). Non-Linear Programming and Stationary Equilibria in Stochastic Games, Tech Rept 87-18, Dept. Math. University of Maryland at Baltimore.

[10] *Fink, A. M.* (1964). Equilibrium Points of Stochastic Noncooperative Games, *J. Sci., Hiroshima Univ., Series A.*, **28**, 89-93.

[11] *Forgess, F* (1986). An Approach to Communication Equilibria, *Econometrica*, **54**, 1375- 1385.

[12] *Frid, E.B.* (1973), On Stochastic Games, *Theory of Probability and its Appl.*, **18**, 389-393.

[13] *Gillette, D.* (1957). Stochastic Games with Zero Stop Probabilities: In Contributions to the Theory of Games Vol III: Edited by M. Dresher A. W. Tucker and P. Wolf, Ann. Math. Studies, **39**, 179-187 Princeton University Press, Princeton, New Jersey.

[14] *Himmelberg, C.J., T. Parthararathy, T.E.S. Raghavan, and F.S. Van Vleck* (1976). Existence of p-Equilibrium and Optimal Stationary Strategies in Stochastic Games, *Proc. Amer. Math. Soc.*, **60**, 245-251.

[15] *Hinderer, K.* (1970). " Foundations of Non-Stationary Dynamic Programming with Discrete Time Parameter", Lecture Notes in Operations Research & Math Systems, **33**, Springer-Verlag, Berlin.

[16] *Hoffman, A.J. and R.M. Karp* (1966). On Non-Terminating Stochastic Games, *Management Science*, **12**, 359- 370.

[17] *Hordijk, A. and L.C.M. Kallenberg* (1981). Linear Programming and Markov Games, in *Game Theory and Mathematical Economics*, Edited by O. Moeschlin and D. Pallaschke, North Holland, Amsterdam, 307-319

[18] *Kohlberg, E.* (1974). Repeated Games with Absorbing States, *Ann. Stat.*, **2**, 724-738.

[19] *Kumar P.R. and T.H. Shiau* (1979). Existence and Value and Randomized Strategies in Zero-Sum Discrete Time Stochastic Dynamic Games, Math. Res. Rept. 79-17, Dec. 1979, University of Maryland at Baltimore County.

[20] *Liggett, T. and S. Lippman* (1969). Stochastic Games with Perfect Information and Time Average Payoff, *SIAM, Rev.*, **11**, 604-607.

[21] *Maitra, A. and T. Parthasarathy* (1970). On Stochastic Games, *J. Opt. Theo. Appl.*, **5**, 289-300.

[22] *Mertens, J.F. and A. Neyman* (1981). Stochastic Games, *Int. J. Game Theory*, **10**, 53-66.

[23] *Mohan, S.R. and T.E.S. Raghavan* (1987). An Algorithm for Discounted Switching Control Stochastic Games, *OR Spektrum*, **9**, 41-45.

[24] *Monash, C.A.* (1979). Stochastic Games, The Minimax Theorem, Ph.D. Thesis, Harvard University, Cambridge, Mass.

[25] *Moulin, H. and J.P. Vial* (1978). Strategically Zero-Sum Games: the Class of Games whose Completely Mixed Equilibria Cannot Be Improved upon, *Int. J. Game Theory*, **7**, 201-221.

[26] *Nowak, A.S.* (1985). Universally Measurable Strategies in Zero-Sum Stochastic Games, *Ann. Prob.*, **13**, 269-287.

[27] *Nowak, A.S.* (1985). Existence of Equilibrium Stationary Strategies in Discounted Non-Cooperative Stochastic Games with Uncountable State Space, *J. Optim. Theory Appl.*, **45**, 591-602.

[28] *Nowak, A.S. and T.E.S. Raghavan* (1988). Existence of Stationary Correlated Equilibria for Discounted Stochastic Games in Arbitrary State Spaces, To appear in *Math. Operations Research.*

[29] *Nowak, A.S. and T.E.S. Raghavan* (1989). A Finite Step Algorithm via a Bimatrix Game to a Single Controller Stochastic Game, Statistical Laboratory Tech. Rept. No. 89-1 February, University of Illinois at Chicago.

[30] *Ornstein, D.* (1969). On the Existence of Stationary Optimal Strategies, *Proc. Amer. Math. Soc.*, **20**, 563-569.

[31] *Parthasarathy, T.* (1982). Existence of Equilibrium Stationary Strategies in Discounted Stochastic Games, *Sankhya* , Ser A. **44**, 114-127.

[32] *Parthasarathy, T. and T.E.S. Raghavan* (1981). An order Field Property for Stochastic Games when One Player Controls Transition Probabilities, *J. Optim. Theory Appl.*, **33**, 375-392.

[33] *Parthasarathy, T., S.H. Tijs and O.J. Vrieze* (1984). Stochastic Games with State Independent Transitions and Separable Rewards. In: Selected Topics in Operation Research and Math.: Edited by H. Moeschlin, and D. Pallaschke, 226-236, Springer-Verlag, Berlin.

[34] *Pollatschek, M. and Avi-Itzhak* (1969). Algorithms for Stochastic Games with Geometrical Interpretation, *Management Sci.*, **15**, 399-425.

[35] *Raghavan, T.E.S, S.H. Tijs, and O.J. Vrieze* (1985). On Stochastic Games with Additive Rewards and Additive Transitions, *J. Optim. Theory Appl.*, **47**, 451-464.

[36] *Raghavan, T.E.S. and J.A. Filar* (1989). Algorithms for Stochastic Games -a Survey, To appear in *Zeits. Operations Research.*

[37] *Rieder, U.* (1979). Equilibrium Plans for Nonzero-Sum Markov Games. In: Game Theory and Related Topics, edited by H. Moeschlin, and D. Pallaschke, North Holland, Amsterdam.

[38] *Rogers, P.* (1969). Nonzero-Sum Stochastic Games, Ph.D. Thesis, University of California, Berkeley, CA.

[39] *Shapley, L.S.* (1953). Stochastic Games, *Proc. Nat. Acad. Sci., U.S.A.*, **39**, 1095-1100.

[40] *Sobel, M.J.* (1971). Non-Cooperative Stochastic Games, *Ann. Math. Stat.*, **42**, 1930-35.

[41] *Sorin, S.* (1980). An Introduction to Two-Person Zero-Sum Repeated Games with Incomplete Information, Tech. Rept. No. 312, The Institute for Mathematical Studies in the Social Sciences, Stanford University, California.

[42] *Stern, M.* (1975). On Stochastic Games with Limiting Average Payoff, Ph.D. Thesis, University of Illinois at Chicago, IL.

[43] *Takahashi, M.* (1964). Equilibrium Points of Stochastic Noncooperative n-Person Games, *J. Sci., Hiroshima Univ. Series, A.*, **28** ,95-99.

[44] *Thuijsman, F. and O.J. Vrieze* (1986). On Equilibria in Repeated Games with Absorbing States Tech. Rept. 8654, Department of Mathematics, Catholic University, Nijmegen, The Netherlands.

[45] *Van der Wal, J.* (1977). Discounted Markov Games: Successive Approximation and Stopping Times, *Internat. J. Game Theory*, **6**, 11-22.

[46] *Vrieze, O.J.* (1981). Linear Programming and Undiscounted Stochastic Game in which One Player Controls Transitions. *OR Spektrum*, **3**, 29-35.

[47] *Vrieze, O.J., S.H. Tijs, T.E.S. Raghavan, and J.A. Filar* (1983). A Finite Algorithm for Switching Control Stochastic Games, *OR Spektrum*, **5**, 15-24.

[48] *Vrieze, O.J.* (1983). Stochastic Games with Finite State and Action Spaces, Ph.D. Thesis, The Free University of Amsterdam, The Netherlands.

[49] *Whitt, W.* (1980). Representation and Approximation of Non-Cooperative Sequential Games, *SIAM J. Control. Optim.*, **18**, 33-48.

A Brief Summary of the Papers in the Volume.

T.E.S. Raghavan

In their **Models for the Game of Liar's Dice**, Ferguson and Ferguson explicitly solve a multimove game of competition where a player must occasionally lie and the other must detect the lie. For example, player I first observes a random variable $X(1)$, having a continuous distribution function $F(x)$. He then chooses $Y(1)$ and claims that $X(1) \geq Y(1)$. Player II, must then challenge or accept player I's claim. If he challenges, player I wins if and only if he was telling the truth. If II accepts the claim, then the game is played again with the roles of the players reversed. Independent of $X(1)$, now player II observes $X(2)$ from $F(x)$, and claims $X(2) \geq Y(2)$ but this time $Y(2)$ must be larger than $Y(1)$. The game may be repeated indefinitely with the players reversing roles and the new call always being greater than the previous call. The value of this game is proved to be $1 - \frac{1}{e}$.

In **A Discrete Search Game**, Ruckle considers the following hide and seek model: A hider hides in one of the locations $1, 2, \ldots, n$. The seeker searches the locations one after the other with returns allowed. He knows that if he visits location i, he will find the hider with chance $0 < p_i \leq 1$, provided that the hider is there. The paper provides a method to approximate the value and also characterizes the best among all pure strategies.

In her paper on **Algorithms for Stochastic Games**, Breton proposes several algorithms for stochastic games with arbitrary transition structure. The algorithms are meant to solve the dynamic programming equations defining equilibria in discounted stochastic games. She also provides the results of an empirical study on some of the existing algorithms.

In their paper **On the Algorithm of Pollatschek and Avi-Itzhak**, Filar and Tolwinski construct a modified Newton-type algorithm for solving discounted zero-sum stochastic games. They show that their modified algorithm does converge to the value vector of the stochastic game from an arbitrary starting point. This resolves the issue raised by a counterexample by Van der Wal [45] against the Pollatschek and Avi-Itzhak algorithm.

In their study of **Semi-infinite Stochastic Games**, Sinha, Thuijsman and Tijs investigate two person zero-sum stopping stochastic games where the action space of player II is countably infinite and everything else is finite. They prove the existence of value, possibly $-\infty$, in some coordinates and show that player I possesses stationary optimals and player II possesses stationary near optimals with finite support.

In their paper on **Easy Initial States in Stochastic Games**, Thuijsman and Vrieze show that for all undiscounted games there exists at least one state for each player, starting from which the player can adopt stationary optimals. In the nonzero-sum case, they exhibit almost-stationary ϵ-equilibria for such starting states.

In many physical models of multistage games, there is a natural asymmetry of information in the sense that one player is well informed about the whole game whereas the opponent might be able to just observe the actions, but not the actual payoffs. In such multistage games with incomplete information, we quote from R.J. Aumann, Survey of Repeated games, in: Essays in Game Theory and Mathematical Economics - in honor of O. Morgenstern, Bibliographisches Institut, Mannheim (1981): "the stress is on the strategic use of information - when and how to reveal and when and how to conceal, when to believe revealed information and when not to, etc. " A survey by Sorin [41], is another accessible source for the major developments in multistage games with incomplete information.

Suppose a stochastic game consists of, say, one of k Big Match type games in the sense of Blackwell and Ferguson [5] and only player I might be well informed, whereas, player II might at best have a prior distribution about the true game being played. In their paper on **"Big Match" with Lack of Information on One Side (III)**, Sorin and Zamir study such games and prove that certain parameterized auxiliary one-shot games $\Gamma(p)$ based on the prior distribution p have minmax value. In **Stochastic ...Positive Stop Probabilities**, Melolidakis shows that the fully informed player has a myopic stationary strategy. The informational difficulties in such games are overcome by an updating mechanism that uses posterior probabilities on the state space given the history up to the present stage.

Moving from finite state to arbitrary states, the intermediate case of countable state stochastic games and special classes such as positive stochastic games generate certain pathologies. For example in a maximization problem of Markovian Decisions with a countable number of states and with finitely many actions, it was proved by Ornstein [30] that when the immediate rewards are nonnegative and the value finite, the optimizer has ϵ-optimal pure stationary strategies.

In the paper on **Positive Stochastic Games and a Theorem of Ornstein**, Nowak and Raghavan construct a countable state positive stochastic game with at most two actions in each state, such that the maximal payoff under any strategies is unity. Unlike in Markovian Decision Theory, the maximizer has no ϵ-optimal stationary strategy. The example is actually a recursive game in the sense of Everett [6] and shows the difficulty in extending the main theorem of Everett to countable state spaces.

The problem of extending stochastic game models to arbitrary state spaces and infinite action spaces, is closely tied to the problems of measurable selections and the analysis of set valued maps. In the case of nonzero-sum games, one could think of extending various equilibrium concepts in the context of stochastic games. In a normal form non-cooperative N-person game, a referee could implement each player's wish in various ways. Let $p_1^*, p_2^*, \ldots, p_N^*$ be probability distributions on the action spaces $A_1, \ldots, A_n$ of the respective players. A referee can select an action according to p_i^* and use it as player i's choice. Thus $p^* = p_1^* \times p_2^* \times \ldots \times p_N^*$ induces a probability distribution on the joint set of actions. We call p^* a Nash equilibrium point if $(p_1^*, \ldots, p_N^*)$ is a Nash equilibrium point. A second possibility is to start with a probability distribution p^* on the joint action space $\Pi_{i=1}^N A_i$. The referee can select an N-tuple $(a_1, \ldots, a_N)$ using p^* and reveal only a_i to player i. Knowing p^* and knowing a_i, if players i, $i = 1, 2, \ldots, N$, have no unilateral gain in altering the action when the rest of the players follow the referee's choice, we call p^* a correlated equilibrium point. The following is another scenario. Before selecting via p^* an action for each player, the referee wants to know whether anyone would like an action different from the one that might be chosen using p^*. Those who prefer to use their own choices must reveal them to the referee. The other players agree to the choice that is dictated by p^*. We call p^* a *correlated weak equilibrium point* if and only if no player in the above situation has any unilateral gain by selecting an action of his own, when the rest of the players agree to the choice according to p^*. Under some normal regularity conditions, Nowak proves the **Existence of Correlated Weak Equilibria in Discounted Stochastic Games with General State Space.**

Since the set of correlated equilibrium payoffs for finite games is a convex set, one could use fixed point arguments to prove the existence of even stationary correlated equilibria for discounted games in arbitrary state spaces [28]. However, the lack of convexity of the set of Nash equilibrium payoffs for finite games prohibits the direct application of fixed point arguments.

In their paper on **Nonzero-Sum Stochastic Games,** Mertens and Parthasarathy show that stochastic games in arbitrary state spaces with compact action spaces admit equilibria in behavior strategies when all transition probabilities are dominated by one single probability measure. They recover a Nash strategy from the convex hull of Nash equilibria of certain dummy games. Here the equilibrium behavior strategy for each player depends only on the current date, current state and the previous state.

Independently, Amir studies the existence of Nash equilibria for discounted stochastic games. In his paper **On Stochastic Games with Uncountable State and Action Spaces,** Amir constructs a Markov equilibrium for the infinite horizon discounted games using the Markov equilibria for finite horizon games, that were obtained earlier by Rieder [37] and Nowak [27]. He assumes that all transitions are dominated by a single probability measure on the state space and the action spaces are finite. When the action spaces are compact, he gives an alternative proof to a theorem of Rieder [37], Nowak [27], and Whitt [49].

From the point of view of some applications in economics, it may be desirable to remove the rigidity of Markovian transitions, typically assumed in models of stochastic games. Rieder studies **Non-Cooperative Dynamic Games with General Utility Functions** where even the payoffs are functions of the entire history. Using the methods of non-stationary dynamic programming initiated by Hinderer [15], Rieder proves the existence of Markov Equilibria when the state space is arbitrary, utility functions are separable and the transitions obey certain regularity conditions.

The concept of equilibria in mixed strategies has been somewhat alien to many practitioners of equilibrium concepts in social sciences. From the point of view of applications, one would like to look for some natural classes of stochastic games where the games admit equilibria in pure stationary strategies.

In their paper on **Symmetric Stochastic Games of Resource Extraction: ...Equilibrium**, Majumdar and Sundaram model a resource extraction problem as a stochastic game. The problem can be stated as follows. Let the stock level of a certain resource commonly shared by two players be s. Both players know the opponent's plan and the stock level. Their actions a_i, to extract a_i units of the resource for $i = 1, 2$ can be carried out if $a_1 + a_2 \leq s$. Otherwise they split the total available resource equally. Unused resource in one period raises to new stock levels in the next period through a renewal function and the realization of a random variable. This defines the transition probabilities. By imposing some natural assumptions on the investment levels and the reproductivity of resources, they derive a pure stationary equilibrium pair for the stochastic game model.

While stochastic games are the natural models for studying multistage games in discrete time, the models for continuous time dynamic games are very many. The independent developments in Control Theory, Differential Games and Dynamical systems try to answer various facets of dynamic optimization and game problems in continuous time. In many situations, one aims to control the behavior of certain physical systems which can be mathematically modeled as a dynamical system. Handling the inherent uncertainties of the physical system is of great importance. By a suitable reformulation of the model, one could reduce the uncertainty problem to the study of stability properties of the new system. In their paper on **Robustness ... Assumptions**, Kai, Leitmann and Garofalo study the robustness of certain controlled systems with delay. Their principal theorem gives an upperbound on the amount of mismatched uncertainty as a function of the chosen control.

Models for the Game of Liar's Dice

Christopher P. Ferguson and Thomas S. Ferguson

Abstract: An explicit multimove game of competition where a player must occasionally lie and the other must detect the lie is solved.

1. Introduction: Models of competition between two decision makers, where one decision maker must occasionally lie and the other must attempt to detect the lie, were described in a nonsequential setting in T. S. Ferguson (1970). Here, several analogous sequential models directly related to liar's dice are treated, in which the decision makers alternately exchange roles. The bigger the successful lie at one stage, the more difficult it is for the opponent at the next stage.

The 1970 model is as follows. Player I observes x chosen from a uniform distribution on $(0,1)$, and claims that he observes $y \in (0,1)$ where $y \geq x$. Player II, not knowing x, is informed of y and must accept or challenge I's claim. If II challenges, I wins 1 from II if $y = x$ and wins nothing otherwise. If II accepts, then I wins $b(x,y)$ from II, where $b(x,y)$ is a known function of x and y. The value and optimal strategies for the players were found for this game under certain monotonicity and differentiability conditions on $b(x,y)$. As examples, the functions $b(x,y) = y$, and $b(x,y) = (y-x)/(1-x)$ were solved explicitly.

No sequential models were treated in Ferguson (1970), although hope was expressed that the results would have application to such models. We quote: "Use of a general $b(x,y)$ allows treatment of situations wherein the basic game is replayed with the roles of the players reversed and future x and y dependent upon past x and y." This may be true in principle, but difficult to carry out in practice.

In this paper, we treat several sequential models directly, all models of the game of liar's dice as described in Bell (1969), Frey (1975) or Scarne (1980). In section 2, a one die version of the game is solved. This is a finite game with at most 6 rounds of play, but with at most 2 rounds under optimal play. In section 3, a continuous version is treated which models liar's dice in which all dice must be rerolled in each round. Although actually an infinite game, under optimal play there is at most 2 rounds. This makes it the game $b(x,y) = y$ in the Ferguson (1970) model, and it has the same solution as given there. In section 4, we treat a version of the game most closely resembling the game of liar's dice as it is played, the seven category game. In section 5, we treat a continuous version modeling liar's dice in which a player may choose which dice to reroll. We call this the continuous improvable case. It is a true stochastic game in the sense of Shapley (1953) (but with continuous state space and some zero stop probabilities) with unbounded length of play under optimal strategies. In section 6, we allow a player to hide from his opponent which of the dice he rerolls. We note that it is advantageous occasionally to try deliberately to get a poor combination of the dice.

2. The Liar's Die Game: As an introductory example for this class of games, we consider the game in which there is just one die. Player I rolls the die and observes the outcome, a random integer $X(1)$ taking the values from 1 to 6 with equal probabilities. Then, based on $X(1)$, player I chooses an integer $y(1)$ between 1 and 6 inclusive and makes the claim that the $X(1)$ is (at least) $y(1)$. Then, player II announces whether he doubts or believes I's claim. If II doubts I's claim, then II wins if $X(1) < y(1)$, and I wins otherwise. If II believes, then the game is played over with the roles of the players reversed. Player II rolls the die but this time II must claim a number that is higher than the number previously claimed by his opponent. Thus, II observes a random integer $X(2)$ between 1 and 6, chooses a number $y(2) > y(1)$, and claims that $X(2) \geq y(2)$. Play proceeds similarly in subsequent rounds. If either player claims 6, then the opponent may as well doubt the claim, since that gives the only possible way to win.

We take the payoff to be 1 if player I wins and 0 if player II wins; this is so that the value of the game will represent player I's probability of win under optimal play by both players. The rules allow the game to be repeated at most six rounds, but under optimal play there are at most two rounds as the following theorem indicates.

Theorem 1: (i) The value of the liar's die game is $41/60$.

(ii) An optimal strategy for player I is as follows. If $X(1) \geq 3$, then claim $y(1) = X(1)$. If $X(1) = 2$, then claim $y(1) = 2$ with probability $3/10$, and claim $y(1) = 3$ otherwise. If $X(1) = 1$, then claim $y(1) = 3$ with probability $3/10$, $y(1) = 4$ with probability $5/10$, and $y(1) = 5$ with probability $2/10$. On the second round (if any), doubt any claim by II.

(iii) An optimal strategy for player II is as follows. If $y(1) = y$, then doubt I's claim with probability $p(y)$, where $p(1) = 0$, $p(2) = 0$, $p(3) = 1/3$, $p(4) = 1/2$, $p(5) = 3/5$, $p(6) = 1$. On the second round (if any), claim the minimum of $X(2)$ and $y(1) + 1$.

Proof: Suppose I uses the indicated strategy. What is II's best reply? If II hears $y(1) = y$, he must doubt if $y = 6$; if $y = 5$, then the probability that $X(1) = 5$ is $P(X(1) = 5|y(1) = 5) = P(X(1) = 5, y(1) = 5)/P(y(1) = 5) = (1/6)/((1/6) + (1/6)(2/10)) = 5/6$; if $y = 4$, then $P(X(1) = 4|y(1) = 4) = 2/3$; if $y = 3$, then $P(X(1) = 3|y(1) = 3) = (1/6)/((1/6) + (1/6)(3/10) + (1/6)(7/10)) = 1/2$.

If II doubts when $y(1) = 5$, he wins with probability 1/6. If II believes, he wins if and only if he rolls a six (since I is going to doubt whatever he does), so here too he wins with probability 1/6. In other words, II is indifferent whether to doubt or believe. Similarly, if $y(1) = 4$, II wins with probability 1/3, whether he doubts or believes, and if $y(1) = 3$, II wins with probability 1/2, whether he doubts or believes. If $y(2) = 2$, clearly II should believe since I is telling the truth, and then since he knows I is going to doubt his claim, which must be at least 3, he should tell the truth if he can and so will win with probability 2/3. All in all, II may as well doubt any $y(1) > 2$ and believe $y(1) = 2$, giving I a probability of win $= 2/3 + (1/6)(3/10)(1/3) = 41/60$, as claimed.

Now suppose II uses the indicated strategy. What is I's best counter strategy? Suppose I sees $X(1) = 1$ or 2; if he claims 3, 4, or 5, he wins with probability $(2/3)(1/2) = 1/3, (1/2)(2/3) = 1/3$, or $(2/5)(5/6) = 1/3$ resp.; if he claims 6, he loses with probability 1; if he claims 2 (1 is worse), he wins with probability 1/3 also. If I sees $X(1) = 3, 4, 5$, or 6, his best policy is honesty, and he wins with probability $(1/3) + (2/3)(1/2) = 2/3, (1/2) + (1/2)(2/3) = 5/6, (3/5) + (2/5)(5/6) = 14/15$, or 1 resp. I's overall probability of winning is therefore $(1/6)(1/3 + 1/3 + 2/3 + 5/6 + 14/15 + 1) = 41/60$, as claimed. ◇

3. The Continuous Liar's Dice Game: Player I observes the outcome of a random variable $X(1)$ having a continuous distribution function, $F(x)$. He then chooses $y(1)$ and claims that $X(1) \geq y(1)$. Player II then must either challenge or accept II's claim. If he challenges, Player I wins if and only if he was telling the truth. If II accepts, then the game is played again with the roles of the players reversed; II observes $X(2)$ from $F(x)$, independent of $X(1)$, and claims $X(2) \geq y(2)$, but this time $y(2)$ must be larger than $y(1)$. The game may be repeated indefinitely, with the players reversing roles, and the new call always greater than the previous call.

A closely related game was analyzed in [3] in which the rules have been altered so that if the second round is reached, player I wins if and only if $X(2) \leq y(1)$. One may think of this as the game in which if the second round is reached, I must challenge II's call. It turns out that this is an optimal strategy for I on the second round so that the value found in [3] is also the value of the game here, namely

$1 - 1/e$. The optimal strategies found there are also optimal first round strategies here, but we give a different (nonrandomized) optimal first round strategy for I in the theorem below.

Since the game depends only on the relative sizes of the $X(n)$ and not on their actual values, we may and do assume without loss of generality that $F(x)$ is the uniform distribution on the interval $[0, 1]$.

Theorem 2: The value of the Continuous Liar's Dice Game is $1 - 1/e$. Optimal for I is the strategy: if $X(1) > 1/e$, then call $y(1) = X(1)$; if $X(1) \leq 1/e$, then call $y(1)$ where $y(1) - 1 - ln(y(1)) = X(1)$; on the second round, challenge any call of II. Optimal for II is the strategy: accept any $y(1) < 1/e$; accept a call of $y(1) > 1/e$ with probability $(1/e)/y(1)$; on the second round, use I's first round strategy except that if it turns out $y(2) < y(1)$, put $y(2)$ anywhere above $y(1)$; on the third round, challenge any call of I.

Proof: Suppose I uses the indicated strategy, and II hears a call of $y(1) = y$. What is the probability I is telling the truth? Let us compute the density of the random quantity $Y = y(1)$ given by I's strategy. This is a change of variable problem with a 2-to-1 transformation. The density of Y is therefore the sum of the two pieces corresponding to $0 < x < 1/e$ and $1/e < x < 1$. The contribution of the first piece is $|dx/dy|$ where $x = y - 1 - ln(y)$, namely, $|1 - 1/y| = 1/y - 1$. The contribution of the second is 1, since the transformation is $x = y$. On hearing $y(1) = y$, the probability that I is telling the truth is therefore the probability that the source is the second contribution, namely, $1/(1 + (1/y - 1)) = y$. If II challenges, then I wins with probability y, but if II accepts, then since I is going to challenge anything II claims, I will win if and only if $X(2) < y$, which also occurs with probability y. Thus it makes no difference what II does against I's strategy; II may as well challenge everything. The probability I wins is then $1 - 1/e$.

Now suppose II uses his indicated strategy, and I observes a value $X(1) = x$. What should I claim? If I claims $y > 1/e$, $y > x$, then I wins iff II accepts and then loses; against II's second round strategy, the best I can do is to challenge any claim of II, so I wins with probability at most $((1/e)/y)y = 1/e$. If $x < 1/e$ and I claims $y > x$ and $y < 1/e$, then II accepts and since on the next round he uses I's original strategy, II wins with probability at least $1 - 1/e$; so again I wins with probability at most $1/e$. If $x < 1/e$ and I calls $y = x$, I wins with probability at most $1/e$. If $x > 1/e$ and I calls $y = x$, then I wins with probability at most $(1-p(y))+p(y)y$ where $p(y) = (1/e)/y$. Since this is greater than $1/e$, I should tell the truth if $x > 1/e$, and it is immaterial whether I lies or tells the truth if $x \leq 1/e$. This strategy holds I's probability of win down to

$$\int_0^{1/e} (1/e)dx + \int_{1/e}^{1} (1 - p(x) + p(x)x)dx = 1 - 1/e.$$

◇

4. Seven-Category Liar's Dice: In this section, we consider liar's dice as it is usually played, allowing a player who accepts a call to leave some of the dice on the table and to roll only those dice that he wishes. The dice not rolled are left in view of his opponent. We consider the realistic case in which there are five dice and to avoid cumbersome details, we restrict attention to seven-category liar's dice, in which if a player accepts a claim by an opponent he must then call a higher category. The seven categories in decreasing order are taken to be 5-of-a-kind, 4-of-a-kind, full house, 3-of-a-kind, two pairs, one pair, and 5 different. Note that a straight is considered as a low hand. Straights are relatively infrequent in games in which it is allowed and it does not pay to try to draw to it. The probabilities of these seven categories are computed in Epstein (1967) and listed in the following table.

Category	# rolls	probability
1. 5 different	720	.0926
2. One pair	3600	.4629
3. Two pairs	1800	.2315
4. 3-of-a-kind	1200	.1543
5. Full house	300	.0386
6. 4-of-a-kind	150	.0193
7. 5-of-a-kind	6	.0008
Total	7776	1.0000

The rules of seven-category liar's dice are as follows. Player I rolls five dice and claims the dice resulted in one of the seven categories above. Player II, not knowing the outcome of the dice must accept or challenge I's claim. If II challenges I's claim, then I wins iff the dice are in the category claimed by I. If II accepts I's claim, the dice are shown to II and the game is repeated with the roles of the players reversed, but this time II may leave some of the dice as they are and roll the rest. The dice not rolled are shown to the opponent and only the new dice are hidden. Now II must claim a higher category than was claimed previously, and it is I's turn to accept or challenge II. Play continues in this fashion until one of the players challenges the other. If a player ever claims the 5-of-a-kind category, it may be assumed that the opponent will challenge, since there is no higher category to call.

This game is less in player I's favor than the game in which all five dice must be rerolled each time a player challenges. It turns out that using optimal strategies, the game never lasts more than three rounds. The critical case occurs when on the first round I truthfully claims category 2 and II accepts the claim. We denote the game continuing from this point by $G(2,2)$ and discuss its solution first.

Suppose then that the last roll resulted in one pair. It is clear that the roller should leave the pair down and roll the remaining three dice. Using such a strategy, the probabilities of improvement to other categories are:

Category	# rolls	probability
2. One pair	60	.2778
3. Two pairs	60	.2778
4. 3-of-a-kind	60	.2778
5. Full house	20	.0926
6. 4-of-a-kind	15	.0694
7. 5-of-a-kind	1	.0046
Total	216	1.0000

The roller should be truthful if he improves, that is, if he rolls one of the categories 3 through 7. If he does not improve, he should claim category y with probability $Q(y)$, where

$$\begin{aligned} Q(6) &= 9/215 = .0419 \\ Q(5) &= 11/100 = .1100 \\ Q(4) &= 8/15 = .5333 \\ Q(3) &= 1 - \text{ the above } = .3148 \end{aligned} \tag{4.1}$$

The main property of this strategy is that the second player is indifferent as to what he should do if he hears a 4, 5 or 6. For example, suppose he hears a 6; then the probability the roller has 4-of-a-kind is $(15/216)/(15/216 + (60/216)(9/215)) = 215/251$. If he challenges, he wins with probability $36/251$; if he accepts, he wins only if he can roll 5-of-a-kind (since his opponent will surely challenge), which happens with probability $1/6$ if he has 4-of-a-kind and with probability $1/216$ if he has a pair, and so with overall probability $(215/251)(1/6) + (36/251)(1/216) = 36/251$. Thus he is indifferent if he hears a 6, and similarly if he hears a 4 or a 5. If he hears a 3, he should accept and, since his opponent will challenge any claim he makes, should lose with probability $120/216$. His overall probability of losing may be computed assuming he challenges 4,5,6 and 7 to be

$$(1 + 15 + 20 + 60)/216 + (120/216)(60/216)(1 + .3148) = .6473\ldots \tag{4.2}$$

The roller can win with this probability at least. To see that this is the value of $G(2,2)$, we find the optimal strategy of the second player. The second player should challenge a claim of y with probability $p(y)$, where $p(2) = p(3) = 0$, $p(7) = 1$, and $p(4)$, $p(5)$ and $p(6)$ are chosen to make the roller indifferent concerning what to call if he rolls a 2:

$$\begin{aligned} (1 - p(4))(180/216) &= (1 - p(5))(200/216) \\ &= (1 - p(6))(215/216) = 120/216 \end{aligned} \tag{4.3}$$

which gives $p(4) = 1/3$, $p(5) = 2/5$, and $p(6) = 19/43$. Since this guarantees the second player that he loses with probability no more than (4.2), this must be the value of the $G(2,2)$.

Theorem 3: The value of seven-category liar's dice is .55919... . I's optimal strategy is as follows. On the first round, if I rolls category 3,4,5,6 or 7, he claims the truth; if I rolls category 1, he claims category 2; if I rolls category 2, he claims category x with probability $P(x)$, where

$$\begin{aligned} P(6) &= 3/430 = .007 \\ P(5) &= 11/400 = .027 \\ P(4) &= 8/45 = .178 \\ P(3) &= 2/5 = .400 \\ P(2) &= 1 - \text{ the above } = .388 \end{aligned}$$

On the second round, I should challenge any claim by II unless I truthfully claimed category 2 in the first round, in which case he should challenge a claim of y by II with the probabilities $p(y)$, associated with (4.3).

II's optimal strategy is as follows. On the first round, II should challenge a claim of y by I with probability $q(y)$ where $q(1) = 0$, $q(2) = 0$, $q(7) = 1$, and $q(3)$ to $q(6)$ are chosen to make I indifferent concerning what to call if he has category 2:

$$\begin{aligned} (1 - q(3))(120/216) &= (1 - q(4))(180/216) \\ &= (1 - q(5))(200/216) \\ &= (1 - q(6))(215/216) = 1 - .6473 \end{aligned}$$

which gives $q(3) = .3651$, $q(4) = .5768$, $q(5) = .6191$, and $q(6) = .6457$. On the second round, II should try to improve the roll except that with two pair he should leave one pair and reroll three dice. Then, II should tell the truth if possible unless he has accepted a truthful claim of 2 or 1. If he has accepted a truthful claim of 2, he should use the Q of (4.1). If he has accepted a truthful claim of 1, he may use I's strategy of the first round, since this never requires a claim of 1.

Sketch of proof: One may check that if I uses the indicated strategy, then II is indifferent as whether to challenge a 3 4 5 or 6. Thus assuming that II challenges 3 through 7, we find that the probability that I wins is $P(\text{I rolls } 1)P(\text{II rolls 1 or 2}) + P(\text{I rolls 2 and calls 2})(1\text{-val}(G(2,2))) + P(\text{I rolls 3 through 7}) = (720/7776)(4320/7776) + (3600/7666)(.388)(1 - .6473) + 3456/7776 = .55919\ldots$. If II uses his indicated strategy, one may also check that I's indicated strategy is optimal against this. ◇

5. Continuous Improvable Liar's Dice: Player I starts as the roller. He observes $X(1)$ from a uniform distribution on the interval $(0,1)$ and claims $y(1) \in (0,1)$. Player II hearing $y(1)$ but not knowing $X(1)$ accepts or challenges. If II challenges, he wins if $X(1) < y(1)$ and loses otherwise. If he accepts, the game is continued with the roles of the players reversed. This time, however, a number $X(2)$ is chosen from the uniform distribution on $(X(1),1)$, and the roller must claim a number $y(2)$ in the interval $(y(1),1)$. This continues *ad infinitum* the game ending after the first challenge, the roles of the players reversing after every acceptance, and $X(n+1) \in \mathcal{U}(X(n),1)$ and $y(n+1) \in (y(n),1)$ for all n. If the game continues forever with neither player challenging, the game is declared a draw although, as we shall see, the probability that this occurs under optimal play of either player is zero. A change of location or scale does not change the problem, so we may as well assume that each $X(n) \in \mathcal{U}(0,1)$ and that the roller must choose $y(n+1)$ from the interval $(z(n),1)$, where $z(n) = (y(n) - X(n))/(1 - X(n))$. Let us denote by $G(z)$ the basic game in which the roller observes $X \in \mathcal{U}(0,1)$ and must choose $y \in (z,1)$. Let $\varphi(z)$ denote the probability that I wins $G(z)$ under optimal play by both players. The original game is $G(0)$, so the value of the original game is $\varphi(0)$. We first present a few lemmas on $\varphi(z)$.

Lemma 1: $\varphi(z)$ is nonincreasing in z, and $\varphi(z) \leq 1 - z$.

Proof: Suppose $z' < z$. Then, any strategy available to I in the game $G(z)$ is also available to I in the game $G(z')$, and it guarantees him the same amount. Hence, $\varphi(z') > \varphi(z)$. Moreover, II can guarantee $1 - z$ by challenging any claim I may make. Then I wins if and only if he tells the truth, which occurs with probability $P(X > z) = 1 - z$. ◇

Lemma 2: Let V and W satisfy

$$V = (1 - V)^{(1+V)/2} = .544658\ldots \tag{5.1}$$
$$W = 1 - V^2/(1 - V) = .348507\ldots$$

Then $\varphi(z) \geq V$ for $0 \leq z \leq W$. Player I can guarantee this amount by using the strategy: if $X = x$, claim $y = f(x)$ where

$$\begin{aligned} f(x) &= [x + V + V^2(1-x)^{-1/V}]/(1+V) \quad \text{for } 0 < x < W \\ &= x \qquad\qquad\qquad\qquad\qquad\qquad\quad\;\; \text{for } W < x < 1. \end{aligned} \tag{5.2}$$

Proof: Suppose $z = 0$. Clearly, II should accept a claim of $y < V$. Suppose II hears a claim of $Y = y > V$ and that $y = f(x)$ where $0 < x < W$. Then the probability that I is telling the truth is $g(y)$, where

$$g(y) = P(X = y|Y = y) = f'(x)/(f'(x) + 1).$$

If II challenges y, then $P(\text{I wins}|Y = y) = g(y)$. If II accepts y, then $P(\text{I wins}|Y = y) = g(y)(1 - \varphi(0)) + (1 - g(y))(1 - \varphi((y - x)/(1 - x)))$. Hence it is optimal for II to challenge y if and only if

$$g(y) \leq g(y)(1 - \varphi(0)) + (1 - g(y))(1 - \varphi((y - x)/(1 - x)))$$

or, equivalently,

$$\varphi(0)f'(x) \leq 1 - \varphi((y - x)/(1 - x)).$$

Suppose $\varphi(0) < V$. Then, using $f'(x) = (f(x) - x)/((1 - x)V)$ for $0 \leq x \leq W$,

$$\begin{aligned}\varphi(0)f'(x) &= \varphi(0)(y - x)/((1 - x)V)\\ &< (y - x)/(1 - x)\\ &\leq 1 - \varphi((y - x)/(1 - x))\end{aligned}$$

from Lemma 1. This implies it is optimal for II to accept a $y < V$ and to challenge a $y > V$, which gives

$$\begin{aligned}P(\text{I wins}) &= (V - W)(1 - \varphi(0)) + (1 - V)\\ &= 1 - W - (V - W)\varphi(0).\\ &= [V^2 - (2V - 1)\varphi(0)]/(1 - V) \qquad (5.3)\\ &> \varphi(0)[V - (2V - 1)]/(1 - V) = \varphi(0).\end{aligned}$$

But this indicates that I can guarantee himself more than the value of the game. This contradiction shows that $\varphi(0) \geq V$. Since I's strategy can be used for $z \leq W$ to guarantee the same amount, we also have $\varphi(z) \geq V$ for $0 \leq z \leq W$. ◇

Lemma 3: $\varphi(z) \geq V$ for $0 \leq z \leq 1 - V$. Player I can guarantee this amount by using the strategy: if $X = x$, claim $y = f(x)$, where

$$\begin{aligned}f(x) &= [x + V + V^2(1 - x)^{-1/V}]/(1 + V) && \text{for } 0 \leq x < W, \qquad (5.4)\\ &= (1 - V)(x - W)/V + 1 - V && \text{for } W < x < 1 - V\\ &= x && \text{for } 1 - V < x < 1.\end{aligned}$$

Proof: Suppose $z = 0$ and II hears a claim of $Y = y$ in $(1 - V, V)$. Then either $Y = y$ or $X = x$ where x is in $(W, 1 - V)$ and $f(x) = y$, the probability of the former being

$$P(X = y|Y = y) = f'(x)/(f'(x) + 1) = 1 - V.$$

Thus, if II accepts I's claim, then

$$P(\text{I wins}|Y=y) = (1-V)(1-\varphi(0)) + V(1-\varphi((y-x)/(1-x)))$$
$$\leq (1-V)(1-V)+V(1-V) = 1-V$$

since $(y-x)/(1-x) = (2V-1)/V = .16398\ldots$, independent of y, and from Lemma 2, $\varphi(0) > V$ and $\varphi(.16398...) > V$. If II challenges, then $P(\text{I wins}|Y = y) = 1-V$ for all y, so II may as well accept all y in $(1-V, V)$. Now follow the rest of the proof of Lemma 2, the only change being that the first line of (5.3) becomes

$$P(\text{I wins}) \geq (V-W)(1-\varphi(0)) + (1-V).$$

Since I's strategy can be used for all $z < 1-V$, we have $\varphi(z) \geq V$ for all z in $(0, 1-V)$. ◇

Theorem 4: The game $G(z)$ has value $\varphi(z) = min\{V, 1-z\}$ where V satisfies (5.1).

Case 1: If $z < 1-V$, each player has an optimal strategy independent of z. The roller has an optimal pure strategy: If $X = x$, claim $y = f(x)$ where f is given by (5.4). (I may better take advantage of poor play by II in $G(z)$ by telling the truth whenever $X > max\{z, W\}$.) An optimal strategy for the challenger is to accept a claim of y with probability $p(y)$, where

$$p(y) = 1 \text{ for } 0 \leq y \leq V, \text{ and} \qquad (5.5)$$
$$= (1-f^{-1}(y))^{1/V} \text{ for } V \leq y \leq 1.$$

Case 2: If $z \geq 1-V$, it is optimal for the challenger to challenge always, and for the roller to claim $y = f(X)$ provided $y > z$, and to claim any $y > z$ otherwise.

Proof: First, suppose $z \geq 1-V$ (Case 2). Player II can guarantee losing at most $1-z$ by challenging always (Lemma 1). On the other hand, if player I claims $y = f(X)$ provided $f(X) > z$, then II would be indifferent between challenging and accepting if it were known that $f(X) > z$; but since there is an additional chance that I is lying, II may as well always challenge. Thus, I is guaranteed at least $1-z$, showing $\varphi(z) = 1-z$.

Next, suppose $z < 1-V$ (Case 1). We have seen (Lemma 3) that the indicated strategy for I guarantees him at least V. Assume then that player II uses the strategy $p(y)$ of (5.5). We complete the proof by showing that II's expected loss is at most V. We find I's best response to $p(y)$.

Suppose I observes $X = x \leq W$. If I lies, he should announce that y in $(V, 1)$ to achieve the maximum of $U(y) = p(y)(1-\varphi((y-x)/(1-x)))$. For $(y-x)/(1-x) > 1-V$, this becomes $U(y) = p(y)(y-x)/(1-x)$. Then since $p'(y) = -p(y)/(y - f^{-1}(y))$, we find that $U'(y) = p(y)(x - f^{-1}(y))/((y - f^{-1}(y))(1 - x))$. Thus among $y \geq x + (1 - x)(1 - V)$, I's return is maximized by $y = f(x)$ and the value of the return there is $U(f(x)) \geq U(f(1)) = 1 - V$. Among $y < x + (1 - x)(1 - V)$, the return is less than $p(y)(1 - V) \leq 1 - V$, so I's optimal response is $y = f(x)$. (Truth also gives $1 - \varphi(0) \leq 1 - V$.)

Suppose I observes $W < x \leq V$. Now lying at $y \geq x + (1 - x)(1 - V)$ is maximized at $y = 1$, with return equal to $1 - V$. Truth also gives at most $1 - V$ as does lying at $y < x + (1 - x)(1 - V)$. So I's optimal response is $y = 1$.

Suppose I observes $V < x \leq 1$. Truth returns $(1 - p(x)) + p(x)(1 - \varphi(0)) = 1 - p(x)\varphi(0)$. Lying should be done at $y = 1$ with return $1 - V$. Thus the optimal return is $max\{1 - V, 1 - p(x)\varphi(0)\}$.

Combining these, the total optimal expected return to I is

$$\int_0^W p(f(x))(f(x)-x)/(1-x)dx+(V-W)(1-V)+\int_V^1 max\{1-V, 1-p(x)\varphi(0)\}dx.$$

Since $(f(x) - x)/(1 - x) = Vf'(x)$, the first term is equal to $V\int_V^1 p(y)dy$. Since $\varphi(0) \geq V$, the third term is less than or equal to $\int_V^1 (1 - p(x)V)dx$. Hence, I's optimal expected return is bounded above by

$$V\int_V^1 p(x)dx+(V-W)(1-V)+\int_V^1 (1-p(x)V)dx = (V-W)(1-V)+(1-V) = V.$$

Since from Lemma 3, I can achieve this amount, II's strategy is optimal. ◇

6. Liar's Coins: In some liar's dice games, the player who accepts a call and rerolls some of the dice is allowed to hide from his opponent which dice are rerolled. This allows the possibility for a player to deliberately try to achieve a low roll by rerolling only those dice that make the roll good in hopes that the opponent will accept the claim and have a hard time improving. We investigate the question of whether this represents a reasonable strategy. To simplify the discussion, we replace poker-type ordering of rolls by the ranking due to the total sum of the dice, and then to simplify further, we replace the dice by coins. The resulting game is as follows.

There are n fair coins and Player I starts by tossing all coins and observing $X(1)$, the total number of heads, and claiming that he sees $y(1)$ heads where $y(1) \geq X(1)$. Player II then challenges or accepts. If II challenges, the game ends in the usual manner. If II accepts, the roles of the players are reversed with II retossing any subset of the coins he wishes. However, II does not show I which coins are retossed. Then, II observes $X(2)$ heads and claims that he sees $y(2)$ heads where $y(2) \geq X(2)$ and $y(2) > y(1)$. Play continues in like manner until one of the players challenges.

First we argue that when n is large it pays occasionally to "go-for-low" by rerolling all of the heads. Except for the first round when all coins must be tossed, the player with control of the coins should either toss all the heads to *destroy* the hand, or all the tails to *improve* it; tossing any other subset of the coins is never of any use. Let x denote the proportion of coins showing heads. By the central limit theorem, if all tails are tossed, the proportion of heads after the toss will be $.5 + x/2 \pm \epsilon$, where ϵ is a term of order $1/\sqrt{n}$. If all heads are tossed, then after the toss there will be a proportion of $x/2 \pm \epsilon$ heads. To simplify things, we take n so large that the ϵ term can be neglected entirely. Assume the controlling player tosses all heads to destroy the hand, and claims $.5 + x/2$. If this claim is believed, the believing player can improve the hand to at most $.5 + x/4$ and will be forced to lie; the other player will win automatically by challenging the claim.

Now, we compute the value, v, of this game, the probability that the controlling player wins under optimal play. The controlling player can either improve the hand or destroy it, in either case claiming $.5 + x/2$. The other player can either believe the claim or challenge it. This leads to a 2×2 matrix game. If the controlling player destroys the hand, he will win if the claim is accepted and lose if it is challenged. If the controlling player improves the hand and is challenged, again he wins. If he improves the hand and is believed, his opponent will win with probability v and so he will win with probability $1 - v$. This results in the following matrix.

	accept	challenge
improve	1 - v	1
destroy	1	0

The value of this game is $1/(1+v)$ which, equated to v, gives $v = (\sqrt{5}-1)/2 = .618\ldots$, the golden ratio, as the value. The optimal strategies are that the controlling player should improve with probability v and destroy with probability $1-v$, and his opponent should accept with probability v and challenge with probability $1-v$. Since the first player must toss all the coins, the second player can believe all but outrageous claims in the first round and become the controlling player for the next round. Thus, under optimal play the first player wins with probability $1-v = .382\ldots$ This is the only game we treat that favors the second player.

Just how many coins are needed to exhibit the phenomenon of going-for-low? The answer is three! Consider the liar's coins game with $n = 3$. Let $P(a, b)$ denote the probability that the player controlling the coins (about to toss any subset) will win under optimal play, given he has a heads and must claim more than b heads. Then, it is straightforward to compute

$$\begin{aligned} P(2,2) &= 1/2 \\ P(1,2) &= 1/4 \\ P(0,2) &= 1/8 \\ P(0,1) &= 1/2. \end{aligned}$$

We now compute $P(1,1)$. The controlling player will claim three heads if and only if he has it, since his opponent is certain to challenge any claim of three heads. Thus, the controlling player has two strategies, improve or destroy, in each case claiming three heads if he has it and two heads otherwise; his opponent has two strategies also, accept a claim of two heads or challenge it. We are led to the matrix game below.

$$\begin{array}{lcc} & \text{accept a two} & \text{challenge} \\ \begin{array}{l}\text{improve}\\ \text{destroy}\end{array} & \multicolumn{2}{c}{\begin{bmatrix} 1/4+(1-P(2,2))/2+(1-P(1,2))/4 & 3/4 \\ (1/2)(1-P(1,2))+(1/2)(1-P(0,2)) & 0 \end{bmatrix}} \end{array}$$

$$= \begin{bmatrix} 11/16 & 3/4 \\ 13/16 & 0 \end{bmatrix}$$

The value of this game is $P(1,1) = 39/56$. The controlling player should improve with probability $13/14$ and destroy with probability $1/14$. His opponent should accept a two with probability $6/7$ and challenge it with probability $1/7$.

Finally we can determine the best strategy for the first round and compute the value of the game. Clearly if the first roll is a two or a three, the first player's claim is the same as his roll. He has a choice of lying schemes for outcomes of 0 and 1 however. He can either claim 1 when he has 0 and sometimes claim 2 when he has 1, or claim 2 when he has 0 and sometimes claim 2 when he has 1. From previous solutions it seems that the first strategy will be better, so we solve using it and hope to show later that any change from it is damaging to his chance of winning. Assuming this strategy, $2/3$ of the time with a 1 he should claim a 2, and the other $1/3$ of the time he should claim a 1. With a 0 he should always claim a 1. The opponent should reject all claims of 3, accept a claim of 2 with probability $17/42$, and accept all claims of 0 or 1. The final step is to demonstrate the validity of our previous assumption on the first player's strategy. This is done by showing that claiming 2 with 0 heads gains nothing over claiming 1 head with 0. The value of this game is $807/1344$.

References

[1] *Bell, R.C.* (1969). "Board and Table Games from Many Civilizations" I, 2nd edition, Oxford University Press.

[2] *Epstein, R.A.* (1967). "The Theory of Gambling and Statistical Logic", Academic Press, New York.

[3] *Ferguson, T.S.* (1970). On a Class of Games Related to Liar Dice, *Ann. Math. Statist.*, **41**, 353-362.

[4] *Frey, Skip* (1977). "Complete Book of Dice Games", Hart Publishing Co. Inc., New York.

[5] *Scarne, John* (1980). " Scarne on Dice", 8th revised edition, Crown Publishers Inc., New York.

[6] *Shapley, L.S.* (1953). Stochastic Games, *Proc. Nat. Acad. Sci. U.S.*, **39**, 1095-1100.

Christopher P. Ferguson and Thomas S. Ferguson
Department of Mathematics
University of California at Los Angeles
Los Angeles, CA 90024
U.S.A.

A Discrete Search Game

William H. Ruckle*

Abstract: A Hider hides in one of the locations $1, 2, \ldots, n$. The Seeker searches the locations one after the other with returns allowed. He knows that if he visits location i, he will find the Hider with chance $0 < p_i \leq 1$, provided that the Hider is there. The paper provides a method to approximate the value and also characterizes the best among all pure strategies.

1. Introduction: In this paper we consider the following search game: the Hider chooses one of the locations $1, 2, \ldots, n$ and remains there. The Seeker searches the locations one after the other with returns allowed, knowing that if he visits location i he will find the Hider with probability $p_i < 1$, provided the Hider is there. The payoff to the Hider is the number of searches the Seeker must make before locating him. If $p_i = 0$ for some i the Hider can obtain an infinite payoff by hiding at location i. In order to avoid this case we assume $p_i > 0$ for each i. In this paper we shall refer to this game as the "Discrete Search Game" or DSG.

Several investigators have found the DSG of interest: e. g. [Bram, 1963], [Neuts, 1963], [Norris, 1962], [Black, 1965], [Dobbie, 1965], [Roberts and Gittins, 1978]. In [Ruckle, 1983] this game is treated on p. 162 ff where it is called ''Hide and Seek on a Complete Graph'' (HSC). The dissertation of Norris cited above considers variations of this game such as the case when the Hider can move but at a cost.

The main purpose of this paper is to establish a sufficient theoretical structure permitting the effective approximation of the solution of the DSG in the sense of Game Theory. That is, we want to approximate the value of the game as well as optimal strategies for Hider and Seeker. The first step in this direction is the

*Research partially supported by NSF DMS 8500946.

mathematical analysis of the DSG implemented by J. Bram in 1963. Next we derive a condition called "The Fundamental Minimality Criterion." This condition provides the desired basis for computation as we illustrate in Section 6. In Section 4 we use Bram's theory and the Fundamental Minimality Criterion to derive the closed form of the solution to the DSG for two locations and $p_i = 1$. In Section 5 we prove that there is a pure strategy in the DSG which may not be optimal but is the best among pure strategies. We give a new derivation of the known solution for the DSG when all p_i are equal and we also show that when each $p_i = .5$ there is an optimal pure strategy for every n.

2. Geometry of the Phase Mapping, The Analysis of J. Bram: A mathematical analysis of the DSG was undertaken by J. Bram in [Bram, 1963]. This analysis provides a good basis for computational solutions to the game. In this section we shall describe the results of this analysis.

A pure strategy for the Seeker in the DSG is an infinite sequence of integers between 1 and n which we shall write as a function f from the integers into the set $\{1, 2, \ldots, n\}$. The set Π of all such functions, that is, the countable product of sets $\{1, 2, \ldots, n\}$ is a compact topological space with the product topology. Convergence in the product topology coincides with pointwise convergence. In other words, the sequence (f_n) converges to f if for each i there is an integer N_i such that $f_n(i) = f(i)$ when $n > N$.

We define the *phase mapping* from Π into R^n by the formula

$$S(f) = b = (b(1), b(2), \ldots, b(n))$$

where $b(i)$ equals the expected payoff to the Hider given that the Hider settles at location i and the Seeker chooses the search sequence f. The i-th coordinate $b(i)$ of $S(f)$ will also be denoted by $S(f)_i$. If the Hider plays the mixed strategy of hiding at location i with probability $u(i)$ and the Seeker plays the pure strategy f then the payoff to the Hider will be $\sum_{i=1}^{n} u(i)S(f)_i$. Here we use the convention $0 \times \infty = 0$. Thus if f and g are two pure strategies for the Seeker such that $S(f) = S(g)$ then the strategies f and g will give the same payoff when played against any pure or mixed strategy of the Hider. This proves that we may regard the Seeker's set of pure strategies as consisting of all points in the set $\{S(f) : f \text{ is a search sequence}\}$. In other words, the Seeker decides upon what $S(f)$ is to be and then chooses an f which will determine that $S(f)$. If $f(m) = i$ for only finitely many m and $p_i < 1$ then $S(f)$ will be ∞ in the i-th coordinate. A search sequence f will be called *admissible* if $S(f)_i$ is finite for each i; i. e. if $S(f)$ is a point in R^n. Let D denote the set of all admissible $S(f)$ and let K denote the closed convex hull of D.

We define the norm $||x||$ of a vector $x = (x(1), x(2), \ldots, x(n))$ to be $max\{|x(i)| : i = 1, 2, \ldots\}$.

If f is a search sequence we define $c(f,i,m)$ to be the number of integers j with $0 \leq j \leq m$ for which $f(j) = i$. If the Seeker employs the search f the expected payoff to the Hider given he chooses the location i will be

$$S(f) = \sum_{m=0}^{\infty} (1-p_i)^{c(f,i,m)} \tag{2.1}$$

because $(1-p_i)_i^{c(f,i,m)}$ is the expected increase of the Hider's payoff at time m. It is not hard to see that the numbers $c(f,i,m)$ satisfy the following conditions

$$\sum_{i=1}^{n} c(f,i,m) = m \tag{2.2}$$

$$\begin{aligned} c(f,i,m+1) &= c(f,i,m) \text{ if } f(m+1) \neq i \\ &= c(f,i,m) + 1 \text{ if } f(m+1) = i. \end{aligned} \tag{2.3}$$

Let f_0 be the search sequence $12\ldots n12\ldots n\ldots$. Then for each i we have

$$S(f_0)_i = \sum_{m=0}^{\infty} (1-p_i)^{c(f,i,m)} = n/p_i - (n-i).$$

Thus f_0 is admissible and $S(f_0) < n/p_i$ for all i. The main result of [Bram, 1963] is the following theorem which provides a geometric characterization of the DSG.

2.1 Theorem: The value of the DSG is $inf\{||x|| : x \in K\}$. An optimal strategy for the Seeker is to choose the point v in K for which all $v(i)$ are equal and $v(i)$ is a minimum. Then $v(i)$ will equal the value of the game. An optimal strategy for the Hider is to hide at location i with probability $u(i)$ where u is any vector normal to a supporting hyperplane of K at v with $\sum_{i=1}^{n} u(i) = 1$.

Theorem 2.1 can be nicely visualized by considering the two dimensional case. The boundary of K forms a curve which is asymptotic to the lines $x = p_1^{-1}$ and $y = p_2^{-1}$. The optimal strategy for the Seeker is a mixture of the two pure strategies (or the pure strategy) at which the line $y = x$ intersects the curve. The common coordinates of the intersection is the value of the game. The optimal strategy for the Hider is (u_1, u_2) which is normal to the curve at the intersection point and satisfies $u_1 + u_2 = 1$. See Figure (6-2). Note that in Figure (6-2) as in all of our other figures we use p for p_1 and q for p_2.

3. A Minimality Criterion: With the help of Bram's theory we shall now develop a criterion for a pure strategy f in Π to minimize $\sum_{j=1}^{n} S(f)_j u_j$ for a given mixed strategy (u_j) for the Hider. In the course of proving Bram's theorem it is shown that for every mixed strategy for the Hider there is a pure strategy of the Seeker which is best against it. Suppose f is optimal against (u_j), $f(i) = x$ and $y \neq x$ is such that $u_x \neq 0$. If $p_y < 1$ there must be a smallest $h > i$ such that $f(h) = y$. Define g to be the search sequence which coincides with f except that $g(i) = y$ and $g(h) = x$. Since f is optimal against (u_j) we must have

$$\sum_{j=1}^{n} u_j S(f)_j \leq \sum_{j=1}^{n} u_j S(g)_j. \tag{3.1}$$

For all j except x and y we have $S(f)_j = S(g)_j$ because for j different from x and y we have $c(f, j, m) = c(g, j, m)$ for each m. Therefore, inequality (3.1) is equivalent to

$$u_x S(f)_x + u_y S(f)_y \leq u_x S(g)_x + u_y S(g)_y$$

or

$$u_y\{S(f)_y - S(g)_y\} \leq u_x\{S(g)_x - S(f)_x\}. \tag{3.2}$$

Because of the definition of $c(f, x, m)$ we see that

$$\begin{aligned} c(g,x,m) &= c(f,x,m) \text{ for } m < i, \\ c(g,x,m) &= c(f,x,m) \text{ for } m \geq h \text{ and} \\ c(g,x,i) &= c(f,x,i) - 1 \text{ if } i \leq m \leq h \end{aligned}$$

It follows that

$$\begin{aligned} S(g)_x - S(f)_x &= (h-i)\{(1-p_x)^{c(f,x,i)-1} - (1-p_x)^{c(f,x,i)}\} \\ &= (h-i)p_x(1-p_x)^{c(f,x,i)-1}. \end{aligned}$$

Similarly we shall have

$$S(f)_y - S(g)_y = (h-i)p_y(1-p_y)^{c(f,y,i)}.$$

Thus we can rewrite equation (3.2) in the form

$$u_y p_y (1-p_y)^{c(f,y,i)} \leq u_x p_x (1-p_x)^{c(f,x,i)-1}. \tag{3.3}$$

Since h is the smallest integer greater than i for which $f(h) = y$ we have $c(f, y, h) = c(f, y, i-1)$; and since $f(i) = x$, $c(f, x, i-1) = c(f, x, i) - 1$. Therefore, (3.3) is equivalent to the equation

$$u_y p_y (1-p_y)^{c(f,y,i-1)} \leq u_x p_x (1-p_x)^{c(f,x,i-1)}. \tag{3.4}$$

Thus we have shown that in order for f to minimize the sum $\sum_{i=1}^{n} u_j S(f)_j$ it is necessary that equation (3.4) be satisfied whenever $f(i) = x$. In other words, it is necessary that for each i, $f(i) = x$ only when $u_x p_x (1 - p_x)^{c(f,x,i-1)}$ is maximal.

We now verify that condition (3.4) is also sufficient to obtain a minimum. Suppose that f and g are two search sequences which satisfy condition (3.4). We shall prove that

$$\sum_{j=1}^{n} u_j S(f)_j = \sum_{j=1}^{n} u_j S(g)_j. \tag{3.5}$$

The number $c(f, x, 1)$ is 1 for some j for which $u_x p_x = u_x p_x (1 - p_x)^{c(f,x,0)}$ is a maximum and the other $c(f, x, j)$ are 0. The numbers $c(g, j, 1)$ are determined in the same way. The number x may not be uniquely determined in this way but the set of numbers $\{u_j p_j (1 - p_j)^{c(f,j,1)} : j = 1, 2, \ldots, n\}$ certainly are. Thus we have

$$\sum_{j=1}^{n} u_j p_j (1 - p_j)^{c(f,j,1)} = \sum_{j=1}^{n} u_j p_j (1 - p_j)^{c(g,j,1)}.$$

Using induction we conclude that for each k the set of numbers $\{u_j p_j (1-p_j)^{c(f,j,k)} : j = 1, 2, \ldots, n\}$ is the same as the set of numbers $\{u_j p_j (1 - p_j)^{c(g,j,k)} : j = 1, 2, \ldots, n\}$. Hence the sum of these numbers are equal. Therefore, we shall have

$$\sum_{k=0}^{\infty} \sum_{j=1}^{n} u_j p_j (1 - p_j)^{c(f,j,k)} = \sum_{k=1}^{\infty} \sum_{j=1}^{n} u_j p_j (1 - p_j)^{c(g,j,k)}.$$

It is not hard to see that the first sum is equal to $\sum_{j=1}^{n} u_i S(f)_j - \sum_{j=1}^{n} u_j (1 - p_j) S(f)_j$ and that the second sum has a corresponding form. Thus we have the equation

$$\sum_{j=1}^{n} u_j S(f)_j - \sum_{j=1}^{n} u_j (1 - p_j) S(f)_j$$

$$\sum_{j=1}^{n} u_j S(g)_j - \sum_{j=1}^{n} u_j (1 - p_j) S(g)_j$$

from which we can derive the equation

$$\sum_{j=1}^{n} u_j \{S(f)_j - S(g)_j\} = \sum_{j=1}^{n} u_j (1 - p_j) \{S(f)_j - S(g)_j\}.$$

But if even one $p_j > 0$ this last equation cannot hold unless both sides are 0 for otherwise the right hand side will be strictly smaller than the left. This implies equation (3.5). Thus we have proved the following result which we shall refer to as the Fundamental Minimality Criterion.

3.1 Theorem: The pure strategy f makes the sum $\sum_{j=1}^{n} u_j S(f)_j$ a minimum if and only if the following condition holds:

For each i we have $f(i) = x$ only if $u_x p_x(1-p_x)^{c(f,x,i-1)}$ is the largest of the quantities $u_y p_y(1-p_y)^{c(f,y,i-1)}$, $y = 1, 2, \ldots, n$.

To use this criterion in computation it suffices to compare the quantities $u_y p_y(1-p_y)^{c(f,y,i-1)}$ at each i to determine the value of $f(i)$. In case of a tie either point can be chosen.

An important special case of Theorem 3.1 occurs when each u_j is proportional to p_j^{-1}, i. e. when

$$u_j = p_j^{-1}\{p_1^{-1} + p_2^{-1} + \ldots + p_n^{-1}\}. \tag{3.6}$$

In this case we have the following simplification of the Fundamental Minimality Criterion.

3.2 Corollary: In the case when each u_j is given by the formula (3.6), the pure strategy f makes the sum $\sum_{j=1}^{n} u_j S(f)_j$ a minimum if and only if the following condition holds:

For each i we have $f(i) = x$ only if $(1-p_x)^{c(f,x,i-1)}$ is the largest of the quantities $(1-p_y)^{c(f,y,i-1)}$, $y = 1, 2, \ldots, n$.

One reason for the importance of the hiding distribution given by (3.6) is that it is the optimal strategy for the Hider in the case when all of the p_j are equal. The following result is found in [Ruckle, 1983]. Using Bram's Theory and the Fundamental Minimality Criterion we can give a substantially easier proof.

3.3 Theorem: If in the DSG all p_j are equal to p, then

(a) an optimal strategy for the Hider is to hide at point j with probability $1/n$ for each $j = 1, 2, \ldots, n$;

(b) an optimal strategy for the Seeker is to choose one of the following search sequences with probability $1/n$:

$$f_j(i) = i + j - 1(\text{mod } n), \qquad j = 1, 2, \ldots, n. \tag{3.7}$$

(c) the value of the game is then $n/p - (n-1)/2$.

Proof: Conclusion (a) is valid by Theorem 2.6 and the symmetry of the surface of K. To verify conclusions (b) and (c) we first observe that each of search sequences f_j satisfies the Fundamental Minimality Criterion with respect to $u_j = 1/n$ for each j. By direct calculation we find that

$$S(f_1)_1 = 1 + n(1-p) + n(1-p)^2 + \ldots = n/p - (n-1)$$
$$S(f_1)_2 = 2 + n(1-p) + n(1-p)^2 + \ldots = n/p - (n-2)$$
$$S(f_1)_3 = 3 + n(1-p) + n(1-p)^2 + \ldots = n/p - (n-3)$$
$$\ldots$$
$$S(f_1)_n = n + n(1-p) + n(1-p)^2 + \ldots = n/p - 0.$$

Therefore, we shall have

$$\sum_{j=1}^{n}(1/n)S(f_1)_j = (1/n)\{n^2/p - 1 - 2 - \ldots - (n-1)\}$$
$$= n/p - (n-1)/2.$$

A similar calculation will show that $\sum_{j=1}^{n}(1/n)S(f_k)_j = n/p-(n-1)/2 = n/p-(n-1)/2$ for each k. Thus the barycenter of the points $\{S(f_k) : k = 1, 2, \ldots, n\}$ lies on the line $x_1 = x_2 = \ldots = x_n$ in R^n. It also lies on the surface of K since each $S(f_k)$ minimizes the functional $\sum_{j=1}^{n}(1/n)S(f_k)_j$. Therefore, by Theorem 2.1 the barycenter $(1/n)\{S(f_1)+\ldots+S(f_n)\}$ represents an optimal mixed strategy for the Seeker. This verifies (b). The validity of (c) follows from the value we obtained for $\sum_{j=1}^{n}(1/n)S(f_k)_j$ for each k. ◇

Another reason for the interest in the mixed strategy for the Hider given by equation (3.5) is the fact that the pure strategies for the Seeker which are optimal against this strategy forms an extensive face of the convex set K which seems to be rather centrally located. There are at least $n!$ such pure strategies because every permutation of $\{1, 2, \ldots, n\}$ satisfies the criterion of Corollary 3.2 as the first n values of a search sequence. It has been widely suspected that the mixed strategy (3.6) may always be optimal for the Hider, but this conjecture is false as is shown by the example in Figure (2-1) where the optimal strategy for the Hider is to hide at 1 with probability approximately .6238 rather than .6 as is given by (3.6).

In the DSG when each p_i is .5 there are optimal pure strategies; one is the search sequence

$$123\ldots(n-1)n(n-1)(n-2)\ldots 1n(n-1)(n-2)\ldots .$$

That is the Seeker passes through the sequence of locations once in ascending order and, thereafter in descending order. The validity of this assertion can be established by calculating that for this search sequence f we have $S(f)_j = 2n - (n-1)/2$ for each j.

4. A Special Case: In this section we shall consider the DSG when $n = 2$, $p_1 = 1$ and $p_2 = p$. If the Hider is at 1, the Seeker will find him with probability 1 when he searches there. If the Hider is at 2 the Seeker will find him with probability p when he searches there. The following theorem provides a complete solution in this special case. In this case of the DSG the Seeker need only 1 a single time. We shall denote by g_m the search sequence for which $g_m(m) = 1$ while $g_m(j) = 2$ for $j \neq m$.

4.1 Theorem: Let h be the largest integer less than $1/p + (1-p)^{h-1}$ and let

$$a = \{(h+1) - 1/p - (1-p)^h\}/\{1 + p(1-p)^{h-1}\}$$
$$\alpha = 1/\{1 + p(1-p)^{h-1}\}.$$

An optimal strategy for the Hider is to hide at 1 with probability $1 - \alpha$ and at 2 with probability α. An optimal strategy for the Seeker is to use the search sequence g_h with probability a and the search sequence g_{h+1} with probability $1 - a$. The value of the game is then $h + 1 - a$.

Proof: For the search sequence g_j we obviously have $S(g_j)_1$. It is also easy to see that

$$S(g_j)_2 = \sum_{n=0}^{j-1}(1-p)^n + (1-p)^{j-1} + \sum_{n=j}^{\infty}(1-p)^n = 1/p + (1-p)^{j-1}.$$

From this it follows that the points $(S(g_j)_1, S(g_j)_2)$, $j = 1, 2, \ldots$ all lie on the curve $y = 1/p + (1-p)^{x-1}$, $x > 0$. Since the second derivative of the function $1/p + (1-p)^{x-1}$ is positive this curve is concave upward. Thus each of the points $(S(g_j)_1, S(g_j)_2)$ lie on the surface of the convex hull K of the image of S. It is clear that any other search sequence will not determine a point on the surface of K since it would waste time by searching 1 more than one time. Thus by Theorem 2.6 the optimal mixed strategy for the Seeker will be found at the point where the line $x = y$ intersects the boundary of K. In other words it suffices to find a between 0 and 1 and an integer h such that

$$ah + (1-h)\{1/p + (1-p)^h\} = a(h+1) + (1-h)\{1/p + (1-p)^{h-1}\}.$$

The solution of this equality determines the h and a in the statement of the theorem. In order to find α we then determine the normal to the line segment joining $S(g_h)$ to $S(g_{h+1})$. This yields the value for α given above. ◇

If there is a p such that $h = 1/p + (1-p)^{h-1}$ has an integral solution in h, then for this p the game treated here would have a pure optimal search sequence g_h while the Hider's optimal strategy would not be unique. It is not known to us whether there is such a p.

4.2 Example: When $p = 1/2$ the optimal strategy for the Hider is to hide at 1 with probability 1/5. The Seeker should search 1 at time 2 with probability 3/5 and search 1 at time 3 with probability 2/5. The value of the game is 13/5.

5. A Best Pure Strategy: We have seen that in the DSG there is an optimal mixed strategy for the Seeker which occurs at the intersection of K, the convex cover of the image of the phase mapping, and the line $x_1 = x_2 = \ldots = x_n$. In this section we show that there is always a pure strategy on this line which is closest to the boundary of K provided $p_i < 1$ for each i. We shall call this the "best" pure search strategy. In general, this pure strategy will not give a longer search time than the optimal mixed strategy. To begin with we prove a technical result concerning the perturbation of a search sequence.

5.1 Lemma: Suppose in the DSG each $p_i < 1$. Let f be an admissible search sequence and let $\epsilon > 0$, and integers i, j be given. Then there is a search sequence g such that

$$S(f)_i < S(g)_i < S(f)_i + \epsilon,$$
$$S(f)_j > S(g)_j > S(f)_j - \epsilon \text{ and}$$
$$S(f)_k = S(g)_k \text{ for } k \neq i \text{ or } j.$$

Proof: Since each $p_i < 1$ and f is admissible (i. e. $S(f)$ is in R^n) it follows that there are arbitrarily large integers q and r such that $f(q) = i$ and $f(r) = j$. Suppose $f(q) = i$ and $f(r) = j$ where $r > q$ and $f(m)$ is neither i nor j for m between q and r. Define g by $g(m) = f(m)$ for m different from q and r, $g(q) = j$ and $g(r) = i$. Then for k different from i and j we have $c(f,k,m) = c(g,k,m)$ for all $m = 1, 2, \ldots$. If $m \leq q$ or $m > r$ then we have $c(f,i,m) = c(g,i,m)$ and $c(f,j,m) = c(g,j,m)$. When $q < m \leq r$ then $c(g,i,m) = c(f,i,m) - 1 = c(f,i,q)$ and $c(g,j,m) = c(f,j,m) + 1 = c(f,j,r+1)$. Consequently we know,

$$S(g)_k = S(f)_k \text{ for } k \text{ different from } i \text{ or } j,$$

$$\begin{aligned} S(g)_i - S(f)_i &= (1-p_i)^{-1} \sum_{m=q+1}^{r} (1-p_i)^{c(f,i,m)} \\ &\leq (1-p_i)^{-1} \sum_{m=q+1}^{\infty} (1-p_i)^{c(f,i,m)}, \end{aligned} \tag{5.1}$$

$$\begin{aligned} S(g)_j - S(f)_j &= (1-p_j) \sum_{m=q+1}^{r} (1-p_j)^{c(f,j,m)}, \\ &\leq (1-p_j) \sum_{m=q+1}^{\infty} (1-p_j)^{c(f,j,m)}. \end{aligned} \tag{5.2}$$

Since the series (5-1) and (5-2) converge we can make the differences less than ϵ by choosing large enough p. ◇

5.2 Theorem: If in the DSG all $p_i < 1$ there is a search sequence f such that all $S(f)_i$ are equal and $S(f)_1$ is a minimum.

Proof: There exists a positive number M such that the set $\Pi_M = \{g : S(g)_n \leq S(g)_{n-1} \leq \ldots \leq S(g)_1 \leq M\}$ is non-empty. To see this, let $M = S(f_0)_1$ where f_0 is the search sequence defined in Section 2. It is not hard to see that Π_M is compact in the topology of pointwise convergence so there is f in Π_M for which $S(f)_1$ is a minimum.

We shall now verify that for the f found above we must have $S(f)_1 = S(f)_2$. If not let ϵ be less than $S(f)_1 - S(f)_2$. By Lemma 5.1 there is a search sequence g such that $S(g)_i = S(f)_i$ for i different from 1 or 2, $S(g)_1 < S(f)_1 < S(g)_1 + \epsilon/2$ and $S(f)_2 < S(g)_2 < S(f)_2 + \epsilon/2$. Then g is in Π_M since

$$S(g)_1 < S(f)_1 \leq M,$$
$$\text{and } S(g)_2 < S(f)_2 + \epsilon/2 < S(f)_1 - \epsilon/2 < S(g)_1.$$

But this contradicts the minimality of $S(f)_1$ since $S(g)_1$ is smaller. Therefore, we must conclude $S(f)_2 = S(f)_1$.

We can continue this process to verify that $S(f)_1 = S(f)_2 = \ldots = S(f)_n$. ◇

In Section 3 we noted that in the DSG with each $p_i = .5$ there are always optimal pure strategies. The following example shows we cannot expect this to happen even when $n = 2$ and $p_1 = p_2$.

5.3 Example: If in the two point DSG $p_1 = p_2 \geq .8$ then there is no optimal pure strategy.

Proof: Let $p_1 = p_2 = p \geq .8$. By Theorem 3.3 the value of the game is $2/p - .5$. If $j \in \{1, 2\}$ is such that $f(1) \neq j$ for the pure strategy f we must have $S(f)_j > 2 = 2/.8 - .5 \geq 2/p - .5$. Thus, f cannot be an optimal pure strategy. ◇

6. Computational Methods: In this section we shall describe four programs used to compute strategies for the DSG and give examples of their application to specific games. These programs have been combined into an integrated Pascal program called SEARPAK which is listed in Clemson U. Tech Report 521 dated 1986.

The routine SEARCH1 determines an optimal search sequence in response to finding probabilities $p_1, p_2, \ldots, p_n$ and a fixed mixed strategy for the Hider. The program listed will handle up to 10 locations. This routine is based upon the application of the Fundamental Minimality Criterion at each step of the search.

The routine SEARCH2 seeks the best pure search sequence as described in Section 5. The algorithm used is simply to search the location j at stage m for which the partial sum $\sum_{i=1}^{m}(1-p_j)^{c(f,j,i)}$ will be the largest. This routine meets with only partial success as the examples below illustrate.

The routines SEARCH3 and SEARCH4 both apply to the DSG in which $n = 2$. They both use the Fundamental Minimality Criterion in order to approximate the game theoretical solution of such a game. The program SEARCH3 plots points $(S(f)_1, S(f)_2)$ where f is the optimal response to the mixed strategy $(u_f, 1-u_f)$ for the Hider. The user can adjust the number of points plotted, the number of searches made and the scale of the graph. The line $x = y$ is also plotted on the graph and the point of intersection of the graph and the line is approximated. SEARCH4 plots points (u_1, y) where u_1 is the probability that the Hider will hide at location 1 and y is the expected search time provided that the Seeker uses the optimal strategy against $(u_1, 1-u_1)$ as determined by the Fundamental Minimality Criterion. The user can limit the domain of u_1 and the range of y in order to obtain increasingly precise solutions. The best procedure to study a particular DSG is to apply SEARCH3 first and then SEARCH4 as many times as necessary to obtain the solution to the desired precision. Norris [Norris, 1962] used a routine similar to SEARCH4, but instead of the Fundamental Minimality Criterion he sought the most likely location from the Bayesian viewpoint

We now present several examples of the DSG to illustrate the use of the four search routines.

6.1 Example: One of the most basic examples of the DSG is the situation when $n = 2$ and $p_1 = p_2 = .5$. By Theorem 3.7 we know that the value of this game is $2/.5 - .5 = 3.5$, an optimal strategy for the Hider is to choose either point with probability .5 and an optimal strategy for the Seeker is to choose with probability .5 one of the following search sequences: 121212...... or 212121....... It can be verified directly that the following search sequences are optimal pure strategies: 1221212...; 21121212.... The second of these optimal pure strategies was located by the application of SEARCH2. See Figure (6.1).

Search Pattern: 2112121212121212121212121212121212121212121212121212121212121
2121212121212121212121212121212121212121212121212

LOCATION	FINDING PROB.	RESULT
1	0.500	3.500
2	0.500	3.500

Figure (6.1)

The result of SEARCH3 in this version of the DSG is shown in Figure (6.2) and the result of SEARCH4 is shown in Figure (6.3). The greater accuracy of the result of

SEARCH4 is due to the fact that in SEARCH3 there is no control over the mixed strategies for the Hider which will be tested by the program.

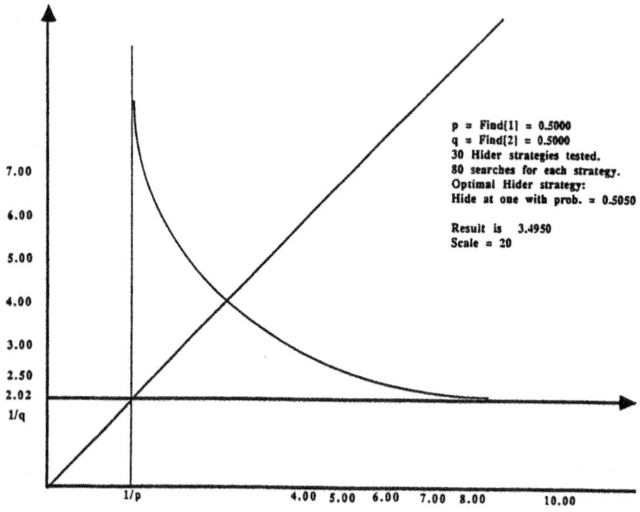

Figure (6.2)

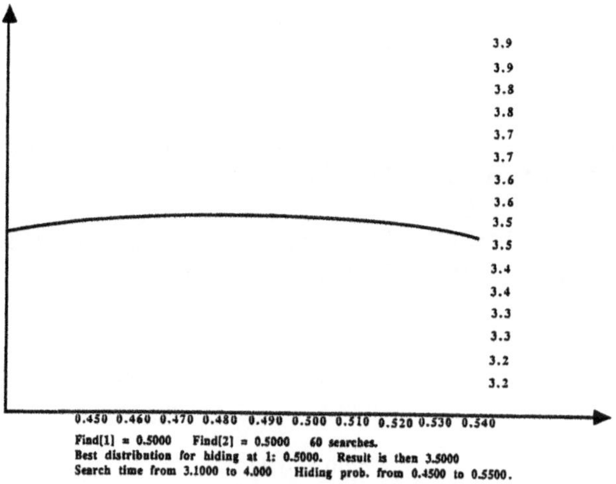

Figure (6.3)

6.2 Example: In Figure (6.4) we illustrate the result of SEARCH1 when $n = 2$, $p_1 = 1$, $p_2 = .5$ and the Hider hides at location 1 with probability .2.

Search pattern: 21222222222222222222

LOCATION	HIDING PROB.	FINDING PROB.	RESULT
1	0.200	1.000	2.000
2	0.800	0.500	2.500

Expected search time is: 2.400.

Figure (6.4)

6.3 Example: Next we treat the case when p_1 is substantially less than p_2 in the two point DSG. Specifically we take $p_1 = .1$ and $p_2 = .4$. The result of SEARCH3 appears in Figure (6.5) and the result of SEARCH4 appears in Figure (6.6).

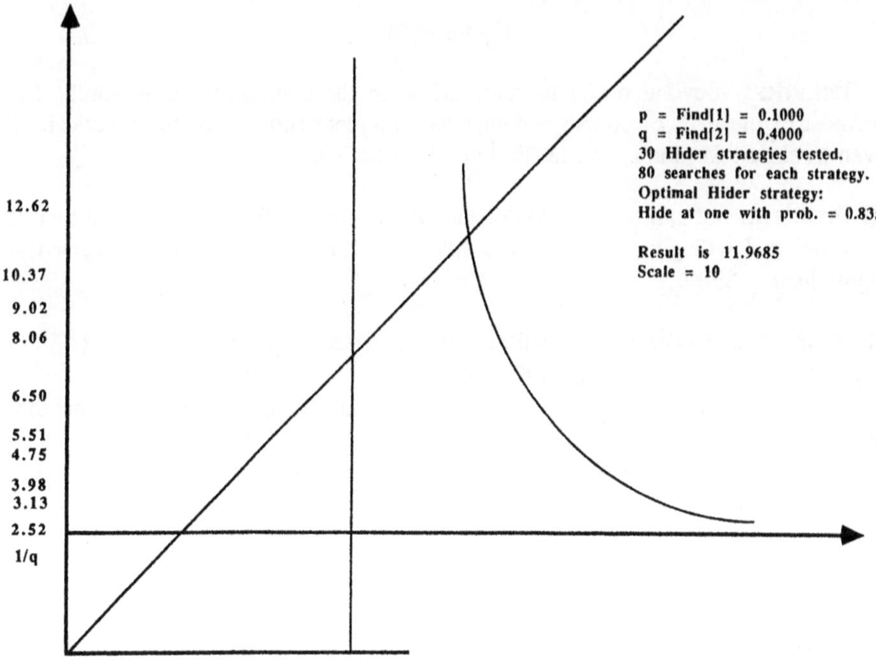

Figure (6.5)

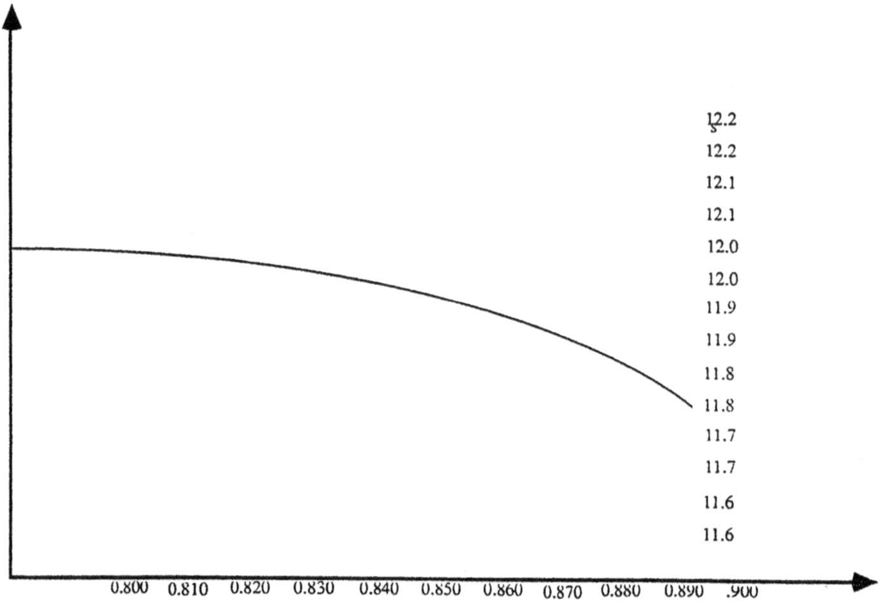

Figure (6.6)

7. Remarks: Several problems have arisen in the course of this research. The foremost of these is, of course, whether the complete solution of the DSG can be given in closed form after the fashion of Theorem 3.3.

7.1 Problem: Obtain the solution of the DSG in closed form. For the solution of this problem I will give the prize of a silk screen print by E. G. Ruckle valued at approximately \$300.

7.2 Problem: Under what conditions is the mixed strategy $u_i = p_i^{-1}/\{p_1^{-1} + \dots p_n^{-1}\}$, $i = 1, 2, \dots, n$, optimal for the Hider.

We have seen that this strategy is optimal when all the p_i are equal. It seems that such a strategy will be optimal when the p_i are "sufficiently close."

We have seen that when each $p_i = .5$ there is an optimal pure search strategy. On the other hand, in Example 5.3 we have seen that even in the two point game with equal p_i there need be no pure search strategy. In this connection we have the following problem:

7.3 Problem: Under what conditions will there be an optimal pure search strategy in the DSG?

For use solutions to Problems 7.2 and 7.3 I offer the prize of a computer game entitled "The Research Empire" published at \$12 and a copy of the book "Geometric Games and their Applications."

References

[1] *Black, W.L.* (1965). Discrete Sequential Search, *Information and Control*, **8**, 159-162.

[2] *Bram, J.* (1963). A 2-Player N-Region Search Game, OEG, ONR, Washington, DC, OEG IRM-31, 21, AD 402 914.

[3] *Dobbie, J.M.* (1974). A Two Cell Model of Search for a Moving Target, *Opns. Res.*, **22**, 79-92.

[4] *Neuts, M.F.* (1963). A Multistage Search Game, *SIAM Journal*, **11**, 502-507.

[5] *Norris, R.C.* (1962). Studies in Search for a Conscious Evader, Dissertation MIT 1962, Lincoln Lab. Tech. Report 279, 134, AD 735-183.

[6] *Roberts, D.M. and Gittins, J.C.* (1978). The Search for an Intelligent Evader: Strategies for Searcher and Evader in the Two Region Problem, *Naval Logistics Quarterly*, 95-106.

[7] *Ruckle, W.H.* (1983). Geometric Games and Their Applications, Boston, London, Melbourne, Pitman.

W.E. Ruckle
Department of Mathematical Sciences
Martin Hall
Clemson University
Clemson, SC 29634-1907
U.S.A.

Algorithms for Stochastic Games*

Michèle Breton

Abstract: In this paper, we present algorithms for the solution of finite discounted stochastic games, without special structure. Three equilibrium concepts are considered: saddle points in two-person zero-sum games, Nash equilibrium points in N-person non-cooperative games and finally Stackelberg equilibrium in two-person games.

Introduction: Stochastic games were introduced by Shapley [13] in which he gave an existence proof of saddle points in zero-sum two-player stochastic games which was constructive and provided a first iterative algorithm for the computation of the value of such games, based on the contraction property of a dynamic programming operator. However, in the general case, even zero-sum stochastic games lack the orderfield property, and therefore successive approximation methods will not converge in a finite number of steps. This motivated interest in finding algorithms for the solution of specially structured games, having the value and data in the same Archimedean ordered field, i.e. single-controller games (Parthasarathy and Raghavan, [8]), switching controller games (Filar, [4], Vrieze, Raghavan, Tijs and Filar, [20]), additive rewards and transition games (Raghavan, Tijs and Vrieze, [12]) and SER-SIT games (Parthasarathy, Tijs and Vrieze, [9]). (See Raghavan, [11] for a survey).

We will present here algorithms for the computation of fixed points of the dynamic programming equations defining equilibrium in discounted stochastic games *without* special structure, in the formalism of Denardo [3], with brief comparisons of their performances on test problems. In the zero-sum case, we propose an efficient iterative algorithm based on results of Pollatschek and Avi-Itzak [10] and Shapley [13] which

*Research supported by NSREC-Canada Grant #A4088. The author wishes to thank Professor Alain Haurie for his helpful comments and suggestions.

will converge to an ϵ-equilibrium from any starting point. In the nonzero-sum N-player case, we propose an implementation of a mathematical program based on results of Schultz [14]. Finally, we propose a finite algorithm for the computation of a Stackelberg equilibrium (min-max) in a zero-sum game, which provides a proof of existence for this solution concept.

Notation: Stochastic games are non-cooperative games played in discrete stages. Let N represent the number of players and S the (finite) set of states of the system. For a given state $s \in S$, let A_{js} represent the (finite) set of admissible actions for player j in state s; $\mathcal{P}(A_{js})$ is the set of all probability distributions on A_{js}. Each player is at each stage of the game fully aware of each action played by the players in the past and of each state occurred in the past (complete information and perfect recall). At each stage, player j observes the state $s \in S$ of the system and chooses, independently from the others, a randomized (or mixed) action α_j in $\mathcal{P}(A_{js})$. An immediate reward $r(j, s, \boldsymbol{\alpha})$ is then earned by player j at this stage, where $\boldsymbol{\alpha}$ represents the vector of (mixed) actions of the N players. The system then moves to a new state s', according to the transition probability $q(s'|s, \boldsymbol{\alpha})$ and the process is repeated from the next stage onward, over the infinite horizon. The players whish to maximize the expected sum of their discounted rewards, where β is the discount factor, $0 < \beta < 1$.

A policy for player j is a function

$$\delta_j : s \in S \rightarrow \delta_j(s) \in \mathcal{P}(A_{js})$$

associating an admissible mixed action to each state of the system. A stationary strategy for player j, denoted also $\boldsymbol{\delta}_j$, consists in using the policy $\boldsymbol{\delta}_j$ at each stage of the game. Let $v_{\boldsymbol{\delta}}(s, j)$ denote the expected discounted sum of the rewards of player j when the initial state of the system is s and when the players use the stationary strategy vector $\boldsymbol{\delta}$.

In order to use the Denardo dynamic programming operators formalism, we associate with each stochastic game a dummy auxiliary one-stage game played as follows: the players observe the state s of the system and choose independently a vector α of mixed actions; each player then receives the immediate reward $r(\cdot, s, \boldsymbol{\alpha})$. A transition occurs to state s', according to the transition probability $q(s'|s, \boldsymbol{\alpha})$, each player receives a terminal reward $v(s', \cdot)$ which depends on the terminal state s' and the game stops. Let $h(s, j, \alpha, v)$ represent the expected reward to player j in this auxiliary dummy game, when the game starts in state s, the players choose the action vector α and when, for a transition to state s', his terminal reward is given by the value $v(s', j)$:

$$h(s, j, \alpha, v) = r(s, j, \alpha) + \beta \sum_{s' \in S} q(s'|s, \alpha) v(s', j). \tag{1}$$

The function h is called the local income function. Let V represent the space of all functions $v : S \times \{1, \ldots, N\} \rightarrow \{R\}$, endowed with the sup norm. The dynamic programming operator $H_\delta : V \rightarrow V$ is defined by

$$(H_\delta v)(s, j) = h(s, j, \delta(s), v). \tag{2}$$

Saddle points in zero-sum stochastic games: In a zero-sum two-player game, the rewards of one player are paid by the other, so that the sum of the rewards of the two players is always zero. In this case, the equilibrium concept is the saddle point: a saddle point in stationary strategies $\delta^* = [\delta_1^*, \delta_2^*]$ is a strategy pair such that, for any stationary strategy pair $[\delta_1, \delta_2]$,

$$v_{[\delta_1, \delta_2^*]}(s, 1) \leq v_{[\delta_1^*, \delta_2^*]}(s, 1) = v^*(s, 1) \leq v_{[\delta_1^*, \delta_2]}(s, 1) \tag{3}$$

and the function v^* is called the value of the game.

Many algorithms were proposed for the computation of saddle points in zero-sum games (for a survey with comparisons on test problems, see Breton, Filar, Haurie and Schultz [2]). For that special case, the "min-max" operator is a contraction mapping; therefore, the value iteration of dynamic programming can be used to compute the value of the stochastic game. This corresponds to the algorithm proposed by Shapley: at iteration n, find the function v^{n+1} satisfying, for all $s \in S$,

$$v^{n+1}(s) = \min_{\delta_2} \max_{\delta_1} (H_\delta v^n)(s). \tag{4}$$

This algorithm requires, at each iteration, the solution of $|S|$ matrix games; it will converge to v^* from any initial v^0.

At iteration n, let

$$DMAX^n = \max v^{n+1}(s) - v^n(s) \; : \; s \in S \tag{5}$$

$$DMIN^n = \min v^{n+1}(s) - v^n(s) \; : \; s \in S. \tag{6}$$

If, for a given $\epsilon > 0$,

$$(DMAX^n - DMIN^n) \leq \epsilon \frac{1-\beta}{\beta} \tag{7}$$

then the policy δ^n satisfying, for all $s \in S$,

$$v^{n+1}(s, 1) = (H_{\delta^n} v^n)(s, 1) \tag{8}$$

also satisfies

$$||v^* - v_{\delta^n}|| \leq \epsilon.$$

However, if at iteration n,

$$||v^{n+1} - v^n|| \leq \epsilon \frac{1-\beta}{\beta} \tag{9}$$

then

$$||v^* - v^n|| \leq \epsilon.$$

In the case of Shapley's algorithm, the choice of stopping criterion (7) instead of (9) will significantly alter the convergence rate. Indeed, the condition (7), which corresponds to finding an ϵ-equilibrium stationary strategy is less restrictive than condition (9), which corresponds to finding a sufficiently close approximation to the value of the game. Using Shapley's algorithm, many iterations are needed to find the value of the game, even after an ϵ-equilibrium stationary strategy has been found. In fact, a value iteration successive approximations algorithm is not an efficient way to compute the reward function $v_{\boldsymbol{\delta}}$ associated with a given stationary strategy vector $\boldsymbol{\delta}$. At iteration n, if condition (7) is satisfied, it suffices to solve the following system of $|S|$ linear equations:

$$v_{\boldsymbol{\delta}^n}(s,1) = (H_{\boldsymbol{\delta}^n} v_{\boldsymbol{\delta}^n})(s,1) \text{ for all } s \in S \tag{10}$$

where $\boldsymbol{\delta}^n$ is a policy vector satisfying (8) in order to obtain a sufficiently close approximation to the value of the game.

Pollatschek and Avi-Itzak [10] proposed a successive iterations algorithm which corresponds to the policy iteration of dynamic programming: At iteration n, find a policy vector $\boldsymbol{\delta}^n$ and the function v^{n+1} such that, for all $s \in S, \delta_1^n$ and δ_2^n are optimal in the matrix games $(H_{\boldsymbol{\delta}^n} v^n), (s,1)$ and

$$v^{n+1}(s,1) = (H_{\boldsymbol{\delta}^n} v^{n+1})(s,1) = v_{\delta^n}(s,1) \tag{11}$$

At each iteration of this algorithm, one needs to solve $|S|$ matrix games and the system of $|S|$ linear equations (11). Van der Wal [18] showed that this algorithm will not converge in general. However, on tests problems, this algorithm performed significantly better than all the others tested, when cycling did not occur.

We now present briefly three other algorithms proposed for the solution of zero-sum stochastic games. Efficiency comparisons of these five methods on test problems can be found in Breton, Filar, Haurie and Schultz [14].

The algorithm of Hoffman and Karp [6] was proposed in order to improve the rate of convergence of Shapley's algorithm: At iteration n, find a policy vector $\boldsymbol{\delta}^n$ satisfying (10) and a function v^{n+1} satisfying, for all $s \in S$,

$$v^{n+1}(s,1) = \max_{\delta_1}(H_{[\delta_1,\delta_2^n]} v^{n+1})(s,1). \tag{12}$$

It was supposed that, since a value iteration algorithm does not use the information contained in the equilibrium policies at each iteration, a policy improvement algorithm would perform better. It was verified on test problems that the Hoffman and Karp algorithm will indeed converge faster than Shapley's algorithm to the value of the game, but in general will not converge faster to the equilibrium strategy. Since each iteration of the Hoffman and Karp algorithm requires more work than an iteration of Shapley's algorithm, it is usually more efficient to use Shapley's algorithm, with the stopping criterion (6), than Hoffman and Karp's algorithm.

Van der Wal [18] proposed an algorithm which is a generalization of both Shapley's and Hoffman and Karp's algorithms; at each iteration, this algorithm finds only an approximate solution to the Markov Decision Process defined in (12): At iteration n, find a policy vector δ^n satisfying (10), then find the next iterate v^{n+1} by applying the operator "max" k times on v^n,

$$v^{n+1}(s,1) = \left(\max_{\delta_1}(H_{[\delta_1,\delta_2^n]}v^n)\right)^k(s,1). \tag{13}$$

One notes that $k = 1$ corresponds to Shapley's algorithm as $k = \infty$ corresponds to Hoffman and Karp's algorithm. On test problems, it was verified that usually this algorithm will converge (from any function $v^0 \geq v^*$) to the value of the game v^* in the same number of iterations that the algorithm of Hoffman and Karp for relatively small values of k (i.e. $k = 3$). This algorithm generally performed better (in terms of CPU times) than Shapley's algorithm in test problems having very sparse transition matrices.

Vrieze and Tijs [19] proposed a different approach to the solution of stochastic games, similar to the fictitious play method for the solution of matrix games. Each iteration of the algorithm corresponds to a stage of a fictitious game; at this stage, each player chooses his action by maximizing his return against the strategy of his opponent corresponding to the frequency distribution of all his past actions. This algorithm only requires at each iteration the computation of maximal and minimal values of the vectors of expected rewards in each state. However, on test problems, the convergence of this algorithm to the value of the game was very slow and moreover, the bounds computed during the process were not tight. The hybrid algorithm that we propose makes use of the efficiency of Pollatschek and Avi-Itzak's and of Shapley's algorithms; moreover, it is guaranteed to converge from any starting function v^0. At iteration n, solve (10) and compute an upper bound ϵ^n on $|v^* - v_{\delta^n}|$:

$$\epsilon^n = \frac{\beta}{1-\beta}(DMAX^n - DMIN^n). \tag{14}$$

As long as the sequence $\{\epsilon^n\}$ is decreasing, redefine v^{n+1} by (11); this corresponds to Pollatschek and Avi-Itzak algorithm. However, if $\epsilon^n \geq \epsilon^{n-1}$, then set $v^{n+1} = v^{n-1}$ and apply Shapley's algorithm from then on. Since Shapley's algorithm will converge from any starting point, this hybrid algorithm will also converge, and will be as efficient as Pollatschek and Avi-Itzak algorithm when cycling of this algorithm does not occur.

With this hybrid algorithm, we have been able to solve in reasonable time stochastic games with up to 20 states, with 20 actions for each player in each state. As a matter of fact, the limitation in size for numerical experiments comes from the memory requirements for the data (transition and rewards matrices).

Nash equilibrium points for N-player stochastic games: In the more general case of non-cooperative stochastic games (N players, general sum), there exist few algorithms for the computation of Nash equilibrium points, even in games with special structure. As a matter of fact, even for *static* non-cooperative games, there exist few efficient methods for the computation of equilibria when there are more than two players.

Consider a stationary strategy vector $\boldsymbol{\delta}$ for the N players and consider, for the jth player, any stationary strategy γ_j. The notation $[\boldsymbol{\delta}^{-j}, \gamma_j]$ represents the stationary strategy vector $\boldsymbol{\delta}'$ such that

$$\delta'_i = \delta_i \text{ for } i \neq j$$
$$\delta'_j = \gamma_j.$$

Let $f_{\boldsymbol{\delta}}$ represent the optimal expected discounted reward of a player when all other players use the given stationary strategy $\boldsymbol{\delta}$; this function is defined at (s, j) by

$$f_{\boldsymbol{\delta}}(s,j) = \max_{\gamma_j}\{v_{[\boldsymbol{\delta}^{*-j},\gamma_j]}(s,j)\}. \tag{15}$$

A stationary strategy vector $\boldsymbol{\delta}^*$ is a Nash equilibrium in stationary strategies if, for all $s \in S$ and $j \in \{1, \ldots, N\}$,

$$v_{\boldsymbol{\delta}^*}(s,j) = f_{\boldsymbol{\delta}^*}(s,j).$$

The existence of a Nash equilibrium in stationary strategies is related to the existence of a fixed point of the point to set mapping Ψ defined by

$$\Psi(\delta)_j = \{\gamma_j \,:\, v_{[\delta^{-j},\gamma_j]}(s,j) = f_{\delta(s,j)} \text{ for all } s \in S\}; \tag{16}$$

A fixed point of Ψ , i.e. a stationary strategy vector $\boldsymbol{\delta}^*$ such that $\boldsymbol{\delta}^* \in \Psi(\boldsymbol{\delta}^*)$ is a Nash equilibrium point. Even for non-cooperative static games, few efficient algorithm exist for the computation of such a fixed point when $N > 2$ (see for example van der Laan and Talman, [17]). In the special case of bi-matrix games (2 players), one can use the algorithm of Lemke and Howson [7]

This algorithm could eventually be used in a successive approximation scheme (value iteration or policy iteration), thus extending Shapley's and Pollatscheck and

Avi-Itzak's algorithms to the nonzero-sum, 2-player case: At iteration n, compute a policy vector δ^n satisfying, for all $s \in S$ and $j \in \{1, 2\}$

$$(H_{\delta^n} v^n)(s, j) \geq (H_{[\delta^{n-j}, \gamma_j]} v^n)(s, j) \text{ for all } \gamma_j. \tag{17}$$

Such a policy vector always exists, since it is a Nash equilibrium point in a bi-matrix game. Then compute the next iterate v^{n+1} satisfying, for all $s \in S$ and $j \in \{1, 2\}$, either

$$v^{n+1}(s, j) = (H_{\delta^n} v^n)(s, j) \tag{18}$$

for a value iteration scheme, or (11) for a policy iteration scheme. However, such methods will not converge in general, since the "equilibrium" operator lacks the monotonicity property. At iteration n, let

$$DMAX^n(j) = max\{(H_{\delta^n} v^n)(s, j) - v^n(s, j) \; : \; s \in S\} \tag{19}$$
$$DMIN^n(j) = min\{(H_{\delta^n} v^n)(s, j) - v^n(s, j) : s \in S\}. \tag{20}$$

It can be proved that

$$||v_{\delta^n} - f_{\delta^n}|| \leq \frac{\beta}{1-\beta} max\Big\{(DMAX^n(j) - DMIN^n(j)) \; : \; j \in \{1, \ldots, N\}\Big\} \tag{21}$$

and a vector δ^n satisfying (17) at iteration n is an ϵ-equilibrium point, where $\epsilon \leq \epsilon^n$ and where

$$\epsilon^n = \frac{\beta}{1-\beta} max\Big\{(DMAX^n(j) - DMIN^n(j)) \; : \; j \in \{1, \ldots, N\}\Big\}. \tag{22}$$

The sequence $\{\epsilon^n\}$ can be used as a stopping criterion: as long as this sequence is decreasing, continue to iterate and stop when $\epsilon^n \geq \epsilon^{n-1}$. In randomly generated test problems, we found that the value iteration algorithm cycled in about half of the cases, and policy iteration cycled even more often. However, these successive approximations methods were successfully used in combination with a mathematical programming approach, as will be described in the sequel.

For the general N-player non-cooperative stochastic game, Breton, Filar, Haurie and Schultz [2] and Schultz [14] proposed a mathematical program to compute a Nash equilibrium point: these authors showed that the global minimum of a polynomial function, subject to linear constraints, corresponds to a Nash equilibrium stationary strategy vector for the stochastic game, and that an approximate solution to this mathematical program corresponds to a Nash ϵ-equilibrium stationary strategy vector. This approach can be related to a feasibility problem proposed by Sobel [16] for the solution of a static game. The mathematical program is:

$$\begin{aligned} min\, f(\mathbf{v},\mathbf{c},\mathbf{p},\mathbf{u},\mathbf{w},\delta) = \sum_{s\in S}\sum_{j=1}^{N}\Big\{ & \left[c_{js} - r_{\delta}(s,j)\right]^2 \\ & + \Big[p_{js} - \sum_{s'\in S} q_{\delta}(s'|s)v_{s'j}\Big]^2 \\ & + \sum_{k\in A_{js}} \Big[u_{jks} - \sum_{s'\in S} q_{[\delta^{-j},k]}(s'|s)v_{s'j}\Big]^2 \\ & + \left[w_{jks} - r_{[\delta^{-j},k]}(s,j)\right]^2\Big\} \end{aligned} \tag{23}$$

s.t., for all $s \in S$, all $j \in \{1, \ldots, N\}$ and all $k \in A_{js}$,

$$v_{sj} \geq w_{jks} + \beta u_{jks} \tag{24}$$

$$v_{sj} = c_{js} + \beta p_{js} \tag{25}$$

$$\sum_{k\in A_{js}} \delta_{jks} = 1 \tag{26}$$

$$\delta_{jks} \geq 0. \tag{27}$$

The value of the functions $r_\delta(\cdot), q_\delta(\cdot), r_{[\delta^{-j},k]}(\cdot)$ and $q_{[\delta^{-j},k]}(\cdot)$ is given by (notice that the vector of variables $\boldsymbol{\delta}_s = [\delta_{jks}], j = 1, \ldots, N, k \in A_{js}$ is a policy vector if the constraints (26)-(27) are satisfied):

$$r_\delta(s,j) = \sum_{\mathbf{a}} r(j, s, \mathbf{a} = [k_1, k_2, \ldots, k_N]) \prod_i \delta_{ik_is} \tag{28}$$

$$q_\delta(s'|s) = \sum_{\mathbf{a}} q(s'|s, = \mathbf{a}[k_1, k_2, \ldots, k_N]) \prod_i \delta_{ik_is} \tag{29}$$

$$r_{[\delta^{-j},k]}(s,j) = \sum_{\mathbf{a}:a_j=k} r(j, s, \mathbf{a} = [k_1, \ldots, k_j = k, \ldots, k_N]) \prod_{i\neq j} \delta_{ik_is} \tag{30}$$

$$q_{[\delta^{-j},k]}(s'|s) = \sum_{\mathbf{a}:a_j=k} q(s'|s, \mathbf{a} = [k_1, \ldots, k_j = k, \ldots, k_N]) \prod_{i\neq j} \delta_{ik_is} \tag{31}$$

The global minimum of this mathematical program is zero, and the optimal values of the vector of variables $\boldsymbol{\delta}$ and v correspond respectively to the Nash equilibrium strategy vector and to the vector of expected discounted rewards for the players.

If T represents the total number of admissible actions, i.e.

$$T = \sum_{j=1}^{N} \sum_{s \in S} |A_{js}|,$$

a Nash equilibrium stationary strategy vector can thus be obtained by minimizing a polynomial function of degree $2(N+1)$ in $3(N|S|+T)$ variables, subject to $2N|S|+T$ linear constraints. The objective function is not convex and could admit more than one local optimum. Therefore, the convergence of an optimization algorithm to the global optimum is not guaranteed. In addition, the efficiency of any solution method depends on the optimization algorithm, and on the starting point.

We implemented this mathematical program for $N = 2$ and $N = 3$, using a sequential quadratic programming technique (NPSOL, Gill and Murray, [5]). Numerical experiments show that the objective function is very ill-conditioned; In order to reach the optimum (which is known to be zero), we used perturbation techniques.

In addition, in the two-players case, we used the value and/or policy iterations algorithms to compute the initial point and to "restart" the search algorithm in the mathematical program: as soon as cycling occurs in the successive iteration algorithm, a starting point is computed, using the last approximations for the equilibrium strategies and the rewards functions in order to obtain a point on the zero level curve of the function f. The sequential quadratic programming algorithm is then used from this point, until either a sufficiently close approximation for the equilibrium strategy and rewards is found or the search fails. In the latter case, the successive iteration algorithm is then applied, using the current value of the variables vector v as the initial reward function v^0, and so on.

We have been able to solve, using the mathematical programming approach, three players stochastic games having up to 5 states and 5 actions per state for each player.

Sequential Stackelberg equilibrium in two-player stochastic games: The sequential Stackelberg equilibrium concept for sequential games was defined by Breton, Alj and Haurie [1] and is related to the feedback Stackelberg solution concept for multi-stage games proposed by Simaan and Cruz [15]. This solution concept corresponds to the case where player 1, the "leader", has to choose his action before player 2, the "follower", who can observe the action chosen by the leader before making his own decision. In the zero-sum case, the sequential Stackelberg equilibrium concept corresponds to the "min-max" solution concept; Breton, Alj and Haurie [1] have shown the existence of sequential Stackelberg equilibrium in pure (non-random) strategies in such games under mild conditions (e.g. in the finite stochastic game case).

Define

$$W = \{(s,a) : s \in S, a \in A_{1s}\}.$$

A stationary strategy δ for the leader is a function associating an admissible mixed action to each state of the system:

$$\delta : s \in S \rightarrow \delta(s) \in \mathcal{P} A_{1s}.$$

A non-random stationary strategy d for the leader is a function associating an admissible action to each state of the system:

$$d : s \in S \rightarrow d(s) \in A_{1s}.$$

A stationary strategy γ for the follower is a function associating an admissible mixed action to each pair $(s,a) \in W$:

$$\gamma : (s,a) \in W \rightarrow \gamma(s,a) \in \mathcal{P} A_{2s}.$$

A non-random stationary strategy g for the follower is a function associating an admissible action to each pair $(s,a) \in W$:

$$g : (s,a) \in W \rightarrow g(s,a) \in A_{2s}.$$

We present first a finite algorithm for the computation of pure sequential Stackelberg equilibrium stationary strategies in a stochastic zero-sum game. It is interesting to note that, in the sequential Stackelberg equilibrium concept case, the follower observes the action chosen by the leader (if the leader is using a mixed strategy, the follower observes the outcome of the randomization), and thus has access to more information than the leader. As a consequence, in the zero-sum case, the sequential Stackelberg equilibrium expected reward for the follower is never less than his saddle-point expected reward in a game having the same data.

Let $v^S(s,1)$ represent the sequential Stackelberg equilibrium expected reward to player 1 when the initial state is s. The following algorithm converges to v^S in a finite number of steps, from any initial function $v^0(\cdot,1) \geq v^S(\cdot,1)$ and provides the corresponding non-random sequential Stackelberg equilibrium strategies: At iteration n, compute a non-random policy g^n for player 2 satisfying, for all $(s,a) \in W$,

$$g^n(s,a) = \text{argmin}\ \{h(s,1,a,b,v^n) : b \in A_{2s}\}. \tag{32}$$

Then compute a non-random policy d^n for player 1 and the next iterate v^{n+1} satisfying, for all $s \in S$,

$$v^{n+1}(s,1) = \left(H_{[d^n,g^n]} v^{n+1}\right)(s,1) = \max_d H_{[d,g^n]} v^{n+1}\big)(s,1). \tag{33}$$

The sequence $\{v^n(\cdot,1)\}$ converges monotonously to $v^S(\cdot,1)$. If $v^{n+1} = v^n$, stop; the function $v^n(\cdot,1) = v^S(\cdot,1)$ and the equilibrium stationary strategies consist in using the policies d^n and g^n.

In the nonzero-sum case, the existence of sequential Stackelberg equilibrium strategies is not guaranteed, even for finite stochastic games. Sufficient conditions for the existence of such equilibria are yet to establish. In Breton, Alj and Haurie [1], a correspondence is shown between sequential Stackelberg equilibria and special forms of Nash equilibria (called strong and weak Nash equilibria) in an associated extended game, having twice as much stages as the original game; this associated extended game is a special case of switching controller games (Filar, [4]).

If a sequential Stackelberg equilibrium exists, the mathematical program (23)-(31) can be transformed in order to take advantage of the special structure of this extended game, and this program can be used to compute sequential Stackelberg equilibrium strategies. This mathematical program is:

$$\min f^S(\mathbf{v},\mathbf{y},\mathbf{p},\boldsymbol{\delta},\gamma) = \sum_{s\in S}\sum_{j=1,2}\Big\{\big[v_{sj} - \sum_{a\in A_{1s}} \delta_{as} y_{saj}\big]^2 + \sum_{a\in A_{1s}}\big[p_{jsa} - \sum_{b\in A_{2s}} \gamma_{bsa} \sum_{s'\in S} q(s'|s,a,b) v_{s'j}\big]^2\Big\} \tag{34}$$

subject to, for all $s \in S, a \in A_{1s}$ and $b \in A_{2s}$,

$$y_{saj} = \sum_{b\in A_{2s}} \gamma_{bsa} r(j,s,a,b) + \beta p_{jsa} \tag{35}$$

$$v_{s1} \geq y_{sa1} \tag{36}$$

$$y_{sa2} \geq r(2,s,a,b) + \beta \sum_{s'\in S} q(s'|s,a,b) v_{s'2} \tag{37}$$

$$\delta_{as} \geq 0 \tag{38}$$

$$\gamma_{bsa} \geq 0 \tag{39}$$

$$\sum_{a\in A_{1s}} \delta_{as} = 1 \tag{40}$$

$$\sum_{b\in A_{2s}} \gamma_{bsa} = 1 \tag{41}$$

Notice that the mappings $\boldsymbol{\delta}(s) = \boldsymbol{\delta}_s = [\delta_{as}] : a \in A_{1s}$ and $\gamma(s,a) = \gamma_{sa} = [\gamma_{bsa}] : (s,a) \in W,\ b \in A_{2s}$ define mixed stationary strategies for the players if the variables $\boldsymbol{\delta}$ and γ satisfy the constraints (38)-(41). If there exists a stationary Stackelberg equilibrium pair for the game, then the optimal value of the function f^S

is zero and the Stackelberg equilibrium strategies and rewards are given by (one of) the solution(s) of the mathematical program (34)-(41). This mathematical program requires the minimization of a polynomial function of degree 4 in respectively

$$2|S| + \sum_{s\in S} |A_{1s}|(5 + |A_{2s}|) \text{ and } \sum_{s\in S} |A_{1s}|(3 + |A_{2s}|)$$

variables and constraints. These become quickly very large numbers as the cardinality of S increases.

Conclusions: In this paper, we presented available algorithms for the computation of equilibrium points in stochastic game without special structure. In the zero-sum case, there exist reasonably efficient algorithms, but such is not the case in the general sum N-players case.

References

[1] *Breton, M., A. Alj, and A. Haurie* (1987). Sequential Stackelberg Equilibria in Two-Person Games, to appear in *Journal of Optimization Theory and Applications.*

[2] *Breton, M., J.A. Filar, A. Haurie and T.A. Schultz* (1986). On the Computation of Equilibria in Discounted Dynamic Games, in *Dynamic Games and Applications in Economics*, Lecture Notes on Economics and Mathematical Systems no. 205, T. Basar, ed., Springer-Verlag, Berlin.

[3] *Denardo, E.V.* (1967). Contraction Mappings in the Theory Underlying Dynamic Programming, *SIAM Review*, **9**, 165-177.

[4] *Filar, J.A.* (1981). Order Field Property for Stochastic Games when the Player who Controls Transitions Changes from State to State, *Journal of Optimization Theory and Applications*, **34**, 505-517.

[5] *Gill, P.E., W. Murray, M.A. Saunders and M.H. Wright* (1983). Users' Guide for SOL/NPSOL: A Fortran Package for Non Linear Programming, Technical report SOL 83-12, Stanford University, Stanford, California.

[6] *Hoffman, A. and R. Karp* (1966). On Non-Terminating Stochastic Games, *Management Science*, **12**, 359-370.

[7] *Lemke, C.E. and J.T. Howson* (1964). Equilibrium Points of Bimatrix Games, *SIAM Journal on Applied Mathematics*, **12**, 413-423.

[8] *Parthasarathy, T. and T.E.S. Raghavan* (1981). An Ordered Field Property for Stochastic Games when One Player Controls Transition Probabilities, *Journal of Optimization Theory and Applications*, **33**, 375-392.

[9] *Parthasarathy, T., S.H. Tijs, and O.J. Vrieze* (1984). Stochastic Games with State Independent Transitions and Separable Rewards. In: Selected Topics in *OR and Math. Economics*, G. Hammer and D. Pallaschke (eds.), Springer Verlag Lecture Notes Series, #226.

[10] *Pollatschek, M. and B. Avi-Itzhak* (1969). Algorithms for Stochastic Games with Geometrical Interpretation, *Management Science*, **15**, 399-415.

[11] *Raghavan, T.E.S.* (1984). Algorithms for Stochastic Games, A Survey, Technical report, University of Illinois at Chicago.

[12] *Raghavan, T.E.S., S.H. Tijs and O.J. Vrieze* (1983). On Stochastic Games with Additive Reward and Transition Structure, Technical report 8339, Mathematics Institute, Catholic University, The Netherlands.

[13] *Shapley, L.S.* (1953). Stochastic Games, *Proceedings of the National Academy of Sciences of USA*, **39**, 1095-1100.

[14] *Schultz, T.A.* (1986). Mathematical Programming and Stochastic Games, Ph.D. Thesis, The Johns Hopkins University, Baltimore, Maryland.

[15] *Simaan, M. and J.B. Cruz Jr.* (1973). On the Stackelberg Strategy in Nonzero-sum Games, *Journal of Optimization Theory and Applications*, **11**, no. 5, 533-555.

[16] *Sobel, M.J.* (1970). An Algorithm for a Game Equilibrium Point, Discussion Paper no. 7035, CORE, Université Catholique de Louvain, Belgique.

[17] *Van der Laan, G. and J.J. Talman* (1982). On the Computation of Fixed Points in the Product Space of Unit Simplices and an Application to Non-cooperative *N* Person Games, *Mathematics of Operations Research*, **7**, no. 1, 1-13.

[18] *Van der Wal, J.* (1978). Discounted Markov Games : Generalized Policy Iteration Method, *Journal of Optimization Theory and Applications*, **25**, 125-138.

[19] *Vrieze, O.J. and S.H. Tijs* (1982). Fictitious Play Applied to Sequences of Games and Discounted Stochastic Games, *International Journal of Game Theory*, **11**, 71-85.

[20] *Vrieze, O.J., T.E.S. Raghavan, S.H. Tijs and J.A. Filar* (1983). A Finite Algorithm for the Switching Control Stochastic Game, *O.R. Spektrum*, **5**, 15-24.

Michèle Breton
GERAD, Ecole des Hautes
Etudes Commerciales de Montréal
Canada.

On the Algorithm of Pollatschek and Avi-Itzhak

Jerzy A. Filar* **and Boleslaw Tolwinski**

Abstract: We present a modification of the Pollatschek and Avi-Itzhak's algorithm for solving the discounted (and terminating), zero-sum, stochastic games. We call our algorithm the Modified Newton's Method and demonstrate that it always converges to the value-vector of the stochastic game, and from an arbitrary starting point. The step-size in our method is selected according to the well-known Armijo's Rule.

1. Introduction: In 1969, in a very interesting paper, Pollatschek and Avi-Itzhak [5] proposed an algorithm for solving zero-sum terminating stochastic games that involved iterations in the policy space, and was closely related to the classical Newton-Raphson procedure. Unfortunately, the above method was proved to be convergent only under rather stringent conditions (see Theorem 5 in [5]).

Nonetheless, the algorithm seemed to perform exceedingly well even in examples that did not satisfy these conditions. Consequently, an attempt was made in Rao et al. [7] to prove that the algorithm always converged. This issue was settled by Van der Wal [10] who gave a simple numerical example in which the Pollatschek and Avi-Itzhak algorithm failed to converge.

The characterization of the class of games for which the above algorithm converges, was posed as an open problem by Van der Wal [10]. However, it could be argued that a more natural problem is to find a modification of this algorithm that will always converge. This point of view is supported by the experience of researchers in nonlinear programming who have discovered a number of successful modifications of Newton's method. In any case, the latter approach is the one adopted in this paper. In particular, in Section 2 we reformulate the fixed-point problem associated with the stochastic game as an unconstrained nonlinear program and demonstrate that a

*Supported in part by the AFOSR and the NSF under the grant ECS-8704954.

Modified Newton's Method (MNM, for short) leads to an algorithm that converges to a global minimum, and from an arbitrary starting point. In Section 3 we demonstrate that our MNM algorithm solves Van der Wal's oscillating example (see [10]) in only 2 major iterations. While our results are derived only for the discounted stochastic games, they extend naturally to Shapley's original terminating games (see [8]).

The relevance of our approach is underscored by a recent empirical study carried out by Breton (see [1], and [2]). This study indicated that the Pollatschek and Avi-Itzhak algorithm, whenever it converged, significantly outperformed all of the known alternative algorithms. Since our MNM algorithm is very closely related to the Pollatschek and Avi-Itzhak's method, it is our expectation that it too will perform well in practice. In Section 4 we mention some preliminary results of numerical experimentation.

2. Notation and Preliminaries: For our purposes, a finite state, finite action space *stochastic game* is defined by: A *state space* $S = \{1, 2, \ldots, n\}$; for each $s \in S$ finite *action spaces* $A(s) = \{1, 2, \ldots, m_s\}$ for player I and $B(s) = \{1, 2, \ldots, n_s\}$ for player II; A *reward law* $r = \{r(s, i, j) \mid (s, i, j) \in S \times A(s) \times B(s)\}$; and a *law of motion* $p = \{p(t|s, i, j) | (s, i, j, t) \in S \times A(s) \times B(s) \times S\}$.

The game is played in *stages* $\tau = 0, 1, 2, \ldots$. At some stage τ, the players find themselves in a state $s \in S$ and independently choose actions $i \in A(s)$, $j \in B(s)$. Player II pays player I the amount $r(s, i, j)$, and at stage $\tau + 1$ the new state is $t \in S$ with probability $p(t \mid s, i, j)$, where $\sum_t p(t \mid s, i, j) = 1$. Play continues at this new state.

In general, strategies can depend on complete histories of the game up until the current stage. We are concerned, however, with the simpler class of stationary strategies which depend only on the current state s. So, for player I, a stationary strategy $\mathbf{x}$ can be denoted by

$$\mathbf{x} = (\mathbf{x}(1), \mathbf{x}(2), \ldots, \mathbf{x}(n)),$$

where $\mathbf{x}(s) = (x_1(s), x_2(s), \ldots, x_{m_s}(s))$ and $x_i(s)$ denotes the probability that player I chooses action $i \in A(s)$ whenever the game is in state $s \in S$. Similarly, a stationary strategy $\mathbf{y}$ for player II can be defined as a set of n vectors $\mathbf{y}(s)$, each of which is a randomization on $B(s)$. Let X and Y denote the sets of all stationary strategies for players I and II respectively.

Once general strategies $\mathbf{x}$ and $\mathbf{y}$ and an initial state $s \in S$ are fixed, the stream of expected payoffs to player I at stage τ, denoted by $V_\tau(s, \mathbf{x}, \mathbf{y}), \tau = 0, 1, 2, \ldots$, is well defined as is the resulting *discounted game payoff*

$$\Phi(s, \mathbf{x}, \mathbf{y}) = \sum_{\tau=0}^{\infty} \beta^\tau V_\tau(s, \mathbf{x}, \mathbf{y}),$$

where $\beta \in (0, 1)$ is the fixed *discount factor.*

A pair of strategies $(\mathbf{x}^0, \mathbf{y}^0)$ is *optimal* for players I and II if for all $s \in S$

$$\Phi(s, \mathbf{x}, \mathbf{y}^0) \leq \Phi(s, \mathbf{x}^0, \mathbf{y}^0) \leq \Phi(s, \mathbf{x}^0, \mathbf{y}),$$

for all strategies $\mathbf{x}$ and $\mathbf{y}$ of players I and II. In such a case, the number $v^*(s) = \Phi(s, \mathbf{x}^0, \mathbf{y}^0)$ will be called *the value* of the game starting in state s, and $\mathbf{v}^* = (v^*(1), \ldots, v^*(n))^T$ will be *the value vector.*

In his fundamental paper Shapley [8] proved that there exist optimal stationary strategies in a class of stochastic games slightly more general than the discounted stochastic games. The proof in [8] is based on the construction outlined below.

Let $\mathbf{v} = (v(1), v(2), \ldots, v(n))^T$ be arbitrary and define for each $(s, i, j) \in S \times A(s) \times B(s)$, $h(\mathbf{v}, s, i, j) = r(s, i, j) + \beta \sum_{t=1}^{n} p(t \mid s, i, j) v(t)$. We can now define a set of n matrix games, of dimensions $m_s \times n_s$, respectively

$$H(\mathbf{v}, s) = (h(\mathbf{v}, s, i, j)).$$

Let the value of each of these be denoted by

$$T(\mathbf{v}, s) := \text{ val } [H(\mathbf{v}, s)] \tag{2.1}$$

for each $s \in S$, and let $\mathbf{T}(\mathbf{v}) = (T(\mathbf{v}, 1), T(\mathbf{v}, 2), \ldots, T(\mathbf{v}, n))^T$. Evidently, we have now defined a contraction map $\mathbf{T} : \mathbf{R}^n \to \mathbf{R}^n$. Shapley [8] proved that $\mathbf{v}^*$, the value of the stochastic game, is the fixed point of $\mathbf{T}$, or equivalently, that it is the unique root of the equation

$$\Psi(\mathbf{v}) := \mathbf{T}(\mathbf{v}) - \mathbf{v} = 0. \tag{2.2}$$

We shall now require additional notation: Let $\mathbf{x}(\mathbf{v}, s)$ and $\mathbf{y}(\mathbf{v}, s)$ be optimal strategies for players I and II, respectively, in the matrix games $H(\mathbf{v}, s)$, for each $s \in S$. Note that these define the stationary strategy in the stochastic game, $\mathbf{x}(\mathbf{v}) = (\mathbf{x}(\mathbf{v}, 1), \mathbf{x}(\mathbf{v}, 2), \ldots, \mathbf{x}(\mathbf{v}, n))$ for player I, and a similarly defined stationary strategy $\mathbf{y}(\mathbf{v})$ for player II. Further, for any stationary strategy pair $(\mathbf{x}, \mathbf{y}) \in X \times Y$ we shall define the following quantities:

$$r(s, \mathbf{x}, \mathbf{y}) := \sum_i \sum_j r(s, i, j) x_i(s) y_j(s), \; s \in S;$$

$$\mathbf{r}(\mathbf{x}, \mathbf{y}) := (r(1, \mathbf{x}, \mathbf{y}), \ldots, r(n, \mathbf{x}, \mathbf{y}))^T;$$

$$p(t \mid s, \mathbf{x}, \mathbf{y}) := \sum_i \sum_j p(t \mid s, i, j) x_i(s) y_j(s), \; s, t \in S;$$

and the transition matrix of a Markov Chain on the states of the game induced by $(\mathbf{x}, \mathbf{y})$, namely $P(\mathbf{x}, \mathbf{y}) = (p(t \mid s, \mathbf{x}, \mathbf{y}))_{s,t=1}^{n,n}$. It is easy to verify that for each $s \in S$

$$\Phi(s, \mathbf{x}, \mathbf{y}) = [(I - \beta P(\mathbf{x}, \mathbf{y}))^{-1} \mathbf{r}(\mathbf{x}, \mathbf{y})]_s, \tag{2.3}$$

where $[\mathbf{u}]_s$ denotes the $s-th$ entry of a vector $\mathbf{u}$. Let $\Phi(\mathbf{x},\mathbf{y})$ denote the n-dimensional vector whose $s-th$ entry is $\Phi(s,\mathbf{x},\mathbf{y})$. Now, with $(\mathbf{x}(\mathbf{v}),\mathbf{y}(\mathbf{v})) \in X \times Y$ defined as above we can associate the Markov matrix

$$P(\mathbf{v}) := P(\mathbf{x}(\mathbf{v}),\mathbf{y}(\mathbf{v})).$$

Our method depends crucially on the following results established by Pollatschek and Avi-Itzhak [5].

Theorem 2.1:

(i) The partial derivatives of

$$\Psi(\mathbf{v},s) := T(\mathbf{v},s) - \mathbf{v}(s)$$

are unique and continuous in $\mathbf{v}$ almost everywhere, for each $s \in S$.

(ii) If for each $s \in S$, $\mathbf{x}(\mathbf{v},s)$ and $\mathbf{y}(\mathbf{v},s)$ are basic optimal strategies in the matrix game $H(\mathbf{v},s)$, in the sense of Shapley and Snow [9], then whenever it is unique

$$\frac{\partial\Psi(\mathbf{v},s)}{\partial v(t)} = \beta p(t \mid s,\mathbf{x}(v),\mathbf{y}(\mathbf{v})) - \delta(s,t),$$

where $\delta(s,t)$ is the Kronecker delta.

(iii) In view of (ii), the gradient matrix of $\Psi(\mathbf{v})$ satisfies:

$$\Psi'(\mathbf{v}) = \beta P(\mathbf{v}) - I$$

almost everywhere.

Note that since $P(\mathbf{v})$ is a Markov matrix and $\beta \in (0,1)$ it follows that the inverse of the gradient matrix

$$[\Psi'(\mathbf{v})]^{-1} = -[I - \beta P(\mathbf{v})]^{-1} \tag{2.4}$$

is well-defined almost everywhere.

Consequently, the iterative step of the classical Newton's method for the solution of (2.2) can be written, with the help of (2.1) and (2.3) as

$$\begin{aligned}
\mathbf{v}^{k+1} &= \mathbf{v}^k - [\Psi'(\mathbf{v}^k)]^{-1}\Psi(\mathbf{v}^k) \\
&= \mathbf{v}^k + [I - \beta P(\mathbf{v}^k)]^{-1}[\mathrm{T}(\mathbf{v}^k) - \mathbf{v}^k] \\
&= \mathbf{v}^k + [I - \beta P(\mathbf{v}^k)]^{-1}[\mathbf{r}(\mathbf{x}(\mathbf{v}^k),\mathbf{y}(\mathbf{v}^k)) \\
&\qquad + \beta P(\mathbf{v}^k)\mathbf{v}^k - \mathbf{v}^k] \\
&= [I - \beta P(\mathbf{v}^k)]^{-1}\mathbf{r}(\mathbf{x}(\mathbf{v}^k),\mathbf{y}(\mathbf{v}^k)) = \Phi(\mathbf{x}(\mathbf{v}^k),\mathbf{y}(\mathbf{v}^k)),
\end{aligned} \tag{2.5}$$

where $\mathbf{v}^k$ is the current estimate of the solution of (2.2). The iterative step (2.5) is the basis of the algorithm of Pollatschek and Avi-Itzhak [5].

3. Modified Newton's Method: While the example of Van der Wal [10] shows that the algorithm of Pollatschek and Avi-Itzhak [5] based on (2.5) need not converge, empirical evidence (see [1], [2]) suggests that it is the fastest known algorithm when it does converge. However, the classical Newton's method for an unconstrained minimization problem can be modified to guarantee that a limit point of a convergent subsequence of estimates is indeed a local minimum. Consequently, in this section we convert the equation (2.2) to a minimization problem, and demonstrate that an analogous modification of (2.5) leads to a globally convergent algorithm to the value vector of the game.

Towards this end we define

$$J(\mathbf{v}) := \frac{1}{2}[\Psi(\mathbf{v})]^T \Psi(\mathbf{v}) \tag{3.1}$$

for all $\mathbf{v} \in R^n$, and consider the unconstrained minimization problem:

$$min[J(\mathbf{v})]. \tag{3.2}$$

Note that the gradient (row) vector of $J(\mathbf{v})$ is given by

$$J'(\mathbf{v}) = [\Psi(\mathbf{v})]^T [\Psi'(\mathbf{v})], \tag{3.3}$$

and that $J'(\mathbf{v}) = 0$ if and only if $\Psi(\mathbf{v}) = 0$ (see Theorem 2.1 (iii)).

We now propose to modify the iterative step (2.5) according to

$$\mathbf{v}^{k+1} = \mathbf{v}^k - w^k[\Psi'(\mathbf{v}^k)]^{-1}\Psi(\mathbf{v}^k), \tag{3.4}$$

where the "step-size" w^k is selected so as to ensure that

$$J(\mathbf{v}^{k+1}) < J(\mathbf{v}^k). \tag{3.5}$$

A popular rule for the choice of w^k is the "First-Order" Armijo's Rule (e.g., see McCormick [4] p. 134). When the iterative scheme (3.4) is applied to the minimization problem (3.2) the following result can be established as a direct consequence of Theorem [12] on page 137 of McCormick [4].

Theorem 3.1: Let the search direction in (3.4) be denoted by $\mathbf{s}^k = -[\Psi'(\mathbf{v}^k)]^{-1}\Psi(\mathbf{v}^k)$, and suppose that w^k is chosen according to Armijo's Rule. Furthermore assume that the following hold:

(a) There exist scalars $\propto, M > 0$ such that for all $k = 1, 2, \ldots,$ we have $\propto ||J'(\mathbf{v}^k)||_2 \leq ||\mathbf{s}^k||_2 \leq M.$

(b) For all k such that $J'(\mathbf{v}^k) \neq 0$

$$\frac{J'(\mathbf{v}^k) \cdot \mathbf{s}^k}{||J'(\mathbf{v}^k)||_2||\mathbf{s}^k||_2} \leq -\delta < 0.$$

Suppose further that $\bar{\mathbf{v}}$ is an accumulation point of the sequence $\{\mathbf{v}^k\}_{k=1}^{\infty}$ generated by (3.4), then

$$J'(\bar{\mathbf{v}}) = 0.$$

Note that in our problem the iterative step (3.4) is equivalent to

$$\begin{aligned} \mathbf{v}^{k+1} &= \mathbf{v}^k + w^k[I - \beta P(\mathbf{v}^k)]^{-1}\{\mathbf{r}(x(\mathbf{v}^k), \mathbf{y}(\mathbf{v}^k)) \\ &\quad - (I - \beta P(\mathbf{v}^k))\mathbf{v}^k\} \\ &= (1 - w^k)\mathbf{v}^k + w^k\Phi(\mathbf{x}(\mathbf{v}^k), \mathbf{y}(\mathbf{v}^k)), \end{aligned} \tag{3.6}$$

and that

$$J'(\mathbf{v}^k) \cdot \mathbf{s}^k = -||\Psi(\mathbf{v}^k)||_2^2 \leq 0. \tag{3.7}$$

Lemma 3.2: Consider the minimization problem (3.2) induced by the discounted stochastic game, and the modified Newton's method (3.4) (equivalently (3.6)), and let $\mathbf{s}^k$ and w^k be as in Theorem 3.1, then conditions (a) and (b) of Theorem 3.1 also hold.

Proof: We shall use the Euclidean and the maximum matrix norms: $||A||_2$ and $||A||_\infty$, respectively, for an arbitrary $n \times n$ matrix A.

Recall also that, in general,

$$||A||_\infty := \max_i \sum_{j=1}^{n} |a_{ij}| \geq (n)^{-\frac{1}{2}}||A||_2.$$

Consequently, if follows from part (iii) of Theorem 2.1 that

$$||\Psi'(\mathbf{v}^k)||_\infty = \max_s \sum_t |\beta p(t|s, \mathbf{x}(\mathbf{v}^k), \mathbf{y}(\mathbf{v}^k)) - \delta(s,t)|. \tag{3.8}$$

Furthermore, for every $s \in S$

$$\sum_t |\beta p(t|s, \cdot, \cdot) - \delta(s,t)| = 1 + \beta[1 - 2p(s|s, \cdot, \cdot)]. \tag{3.9}$$

It now follows that for every $k = 1, 2, \ldots$

$$(1-\beta) \leq \|\Psi'(\mathbf{v}^k)\|_\infty \leq (1+\beta), \tag{3.10}$$

and hence that with $c_1 = n^{\frac{1}{2}}$

$$\|\Psi'(\mathbf{v}^k)\|_2 \leq c_1(1+\beta). \tag{3.11}$$

Similarly, with the help of the fact that $P(\mathbf{v}^k)$ is a Markov matrix, we obtain

$$\|\beta^\tau P^\tau(\mathbf{v}^k)\|_\infty = \beta^\tau \|P^\tau(\mathbf{v}^k)\|_\infty = \beta^\tau < 1, \tag{3.12}$$

for every $k = 1, 2, \ldots$, and every $\tau = 0, 1, 2, \ldots$. It now follows that

$$\|[I - \beta P(\mathbf{v}^k)]^{-1}\|_\infty \leq \sum_{\tau=0}^{\infty} \beta^\tau = \frac{1}{1-\beta}, \tag{3.13}$$

and hence that for every $k = 1, 2, \ldots$

$$\|[\Psi'(\mathbf{v}^k)]^{-1}\|_2 = \|[I - \beta P(\mathbf{v}^k)]^{-1}\|_2 \leq \frac{c_1}{1-\beta}. \tag{3.14}$$

From the definitions of $\mathbf{s}^k$ and $J'(\mathbf{v}^k)$ we now obtain, with the help of (3.7), (3.11) and (3.14), that

$$\begin{aligned}
\frac{J'(\mathbf{v}^k) \cdot \mathbf{s}^k}{\|J'(\mathbf{v}^k)\|_2 \|\mathbf{s}^k\|_2} &= -\frac{\|\Psi(\mathbf{v}^k)\|_2^2}{\|J'(\mathbf{v}_k)\|_2 \|\mathbf{s}^k\|_2} \\
&\leq -\frac{\|\Psi(\mathbf{v}^k)\|_2^2}{\|\Psi(\mathbf{v}^k)\|_2 \|\Psi'(\mathbf{v}^k)\|_2 \|[\Psi'(\mathbf{v}^k)]^{-1}\|_2 \|\Psi(\mathbf{v}^k)\|_2} \\
&\leq -\frac{1-\beta}{c_1^2(1+\beta)} = -\delta < 0,
\end{aligned} \tag{3.15}$$

where $\delta = (1-\beta)/c_1^2(1+\beta)$. Consequently, condition (b) of Theorem 3.1 is satisfied.

To verify condition (a) of Theorem 3.1 note that for every $k = 1, 2, \ldots$

$$\begin{aligned}
\|J'(\mathbf{v}^k)\|_2 &= \|[\Psi'(\mathbf{v}^k)]^T \Psi(\mathbf{v}^k)\|_2 \\
&= \|[\Psi'(\mathbf{v}^k)]^T \Psi'(\mathbf{v}^k)[\Psi'(\mathbf{v}^k)]^{-1} \Psi(\mathbf{v}^k)\|_2 \\
&\leq \|[\Psi'(\mathbf{v}^k)]^T\|_2 \|\Psi'(\mathbf{v}^k)\|_2 \|\mathbf{s}^k\|_2.
\end{aligned} \tag{3.16}$$

Similarly to (3.12) it can be checked that for every $k = 1, 2, \ldots$

$$\|[\Psi'(\mathbf{v}^k)]^T\|_2 \leq c_2(1+\beta), \tag{3.17}$$

where $c_2 = n^{\frac{3}{2}}$. Substituting (3.17) and (3.11) into (3.16) we obtain that for every $k = 1, 2, \ldots$

$$\|\mathbf{s}^k\|_2 \geq \frac{1}{c_1 c_2 (1+\beta)^2} \|J'(\mathbf{v}^k)\|_2 \tag{3.18}$$

which completes the proof of this lemma. ◇

Theorem 3.3: Let $M = max|r(s,i,j)|$ and set $\mathbf{v}^1 = M\mathbf{1}$, where $\mathbf{1}$ is an n-dimensional vector with every entry equal to 1. Using $\mathbf{v}^1$ as the initial point generate a sequence $\{\mathbf{v}^k\}_{k=1}^{\infty}$ by the iterative scheme (3.6) where the step-size w^k is selected according to Armijo's rule and $(\mathbf{x}(\mathbf{v}^k), \mathbf{y}(\mathbf{v}^k))$ are made up from basic optimal strategies in the matrix games $H(\mathbf{v}^k, s)$ for each $s \in S$, and for every $k = 2, 3, \ldots$. The sequence $\{\mathbf{v}^k\}_{k=1}^{\infty}$ converges to the limit $\mathbf{v}^*$ which is the value vector of the stochastic game.

Proof: We first note that by Theorem 3.1 and Lemma 3.2 any accumulation point $\bar{\mathbf{v}}$ of the sequence $\{\mathbf{v}^k\}_{k=1}^{\infty}$ satisfies $J'(\bar{\mathbf{v}}) = 0$, and hence also $\Psi(\bar{\mathbf{v}}) = 0$ (see (3.3)). Hence by (2.2), $\bar{\mathbf{v}}$ is the value vector of the stochastic game. By the uniqueness of the latter, the sequence $\{\mathbf{v}^k\}_{k=1}^{\infty}$ can have only one accumulation point. Our proof will be complete once we demonstrate that this sequence is bounded, because then we shall know that one accumulation point $\mathbf{v}^*$ must exist and (by uniqueness of accumulation points) it must be the limit of the original sequence. However, we now note from (3.6) that for every $k = 1, 2, \ldots$

$$\begin{aligned} ||\mathbf{v}^{k+1}||_2 &\leq (1 - w^k)||\mathbf{v}^k||_2 + w^k||\Phi(\mathbf{x}(\mathbf{v}^k), \mathbf{y}(\mathbf{v}^k))||_2 \\ &\leq (1 - w^k)||\mathbf{v}^k||_2 + w^k M/(1-\beta) \end{aligned} \tag{3.19}$$

Further, since $w^k \in [0,1]$ for every k, and since $||\mathbf{v}^1||_2 \leq M$ by construction, we have that

$$||\mathbf{v}^{k+1}||_2 \leq M/(1-\beta), \qquad k = 2, 3, \ldots, \tag{3.20}$$

which completes the proof. ◇

The above results lead to the following globally convergent algorithm for the solution of zero-sum discounted stochastic games.

Algorithm: (Modified Newton's Method, MNM)

Step 0. Set $k := 0$. Select two parameter values: $\alpha \in (0,1)$ ("small" value of α is generally recommended), and $\mu \in [0.5, 0.8]$. Also select $\mathbf{v}^0$, the initial estimate of the value vector.

Step 1. Calculate for each $s \in S$, the matrix game $H(\mathbf{v}^k, s)$, a pair of optimal basic strategies $(\mathbf{x}(\mathbf{v}^k, s), \mathbf{y}(\mathbf{v}^k, s))$ for players I and II (see [3]), and its value $T(\mathbf{v}^k, s)$. Calculate $\mathrm{T}(\mathbf{v}^k)$, $\Psi(\mathbf{v}^k)$ and $J(\mathbf{v}^k)$ (see (2.1), (2.2) and (3.1)).

Step 2. If $J(\mathbf{v}^k) = 0$, stop; $\mathbf{v}^k$ is the value vector of the stochastic game.

Step 3. Calculate $\mathbf{s}^k$ as in Theorem 3.1.

Step 4. Set $w^k := 1$.

Step 5. Test the inequality

$$J(\mathbf{v}^k + w^k\mathbf{s}^k) - J(\mathbf{v}^k) \leq \alpha\, w^k \mathbf{J}'(\mathbf{v}^k)\mathbf{s}^k.$$

If the above inequality is satisfied set $\mathbf{v}^k := \mathbf{v}^k + w^k\mathbf{s}^k$, $\quad k := k+1$ and return to Step 2. Otherwise, go to Step 6.

Step 6. Set $w^k := \mu w^k$ and return to Step 5.

4. The Example of Van der Wal: In an interesting paper, Van der Wal [10] demonstrated by an example that the original scheme (2.5) of Pollatschek and Avi-Itzhak [5] need not converge to the value vector of the stochastic game. In this section we shall apply our modified scheme (3.6) to the same example and demonstrate convergence in two iterations. The data of the example is as follows:

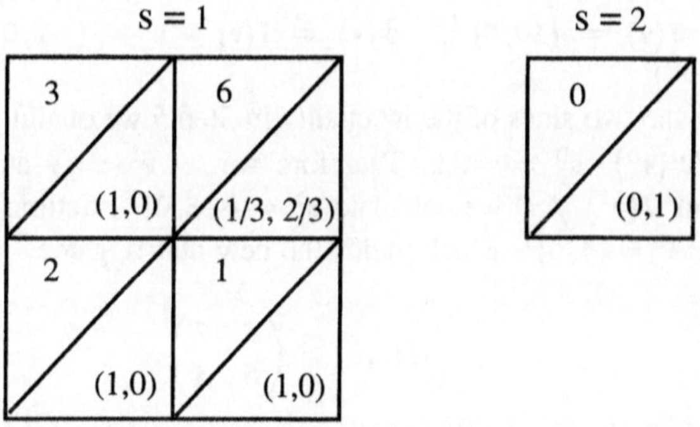

The above notation has this interpretation: The state $s = 2$ is absorbing and carries with it the payoff of 0; the choice in state $s = 1$ of, say, actions 1 and 2 by players I and II respectively results in the payoff of 6 and the transition to state 2 with probability of $\frac{2}{3}$; and so on. We take the discount factor $\beta = \frac{3}{4}$. We shall begin, as in Van der Wal, with $\mathbf{v}^0 = (0,0)^T$ and parameter values $\alpha = .1, \mu = .5$. The matrix game

$$H(\mathbf{v}^0, 1) = \begin{pmatrix} 3 & 6 \\ 2 & 1 \end{pmatrix}$$

has value $T(\mathbf{v}^0, 1) = 3$ and a pair of basic optimal strategies $\mathbf{x}(\mathbf{v}^0, 1) = (1,0) = \mathbf{y}(\mathbf{v}^0, 1)$. Of course, we are ignoring state 2 since both players have only one choice in that state. We now have

$$\mathsf{T}(\mathbf{v}^0) = (3,0)^T = \Psi(\mathbf{v}^0),$$

Since $J(\mathbf{v}^0) = \frac{9}{2} \neq 0$ we proceed with Step 3 of the Modified Newton's Method. It is easy to check that

$$\Psi'(\mathbf{v}^0) = \begin{pmatrix} -\frac{1}{4} & 0 \\ 0 & -\frac{1}{4} \end{pmatrix},$$

$\mathsf{J}'(\mathsf{v}^0) = (-\frac{3}{4}, 0)$ and $\mathbf{s}^0 = (12, 0)^T$. We now set $w^0 = 1, \mathsf{v} := \mathsf{v}^0 + w^0\mathbf{s}^0 = (12, 0)^T$, and calculate the matrix game

$$H(\mathsf{v}, 1) = \begin{pmatrix} 12 & 9 \\ 11 & 10 \end{pmatrix}$$

which has value $T(\mathsf{v}, 1) = 10$ and optimal basic strategies $\mathsf{x}(\mathsf{v}, 1) = \mathsf{y}(\mathsf{v}, 1) = (0, 1)$. These yield $\mathsf{T}(\mathsf{v}) = (10, 0)^T$, $\Psi(\mathsf{v}) = \mathsf{T}(\mathsf{v}) - \mathsf{v} = (-2, 0)^T, J(\mathsf{v}) = 2$, and $\Psi'(\mathsf{v}) = \Psi'(\mathsf{v}^0)$.

Evaluating the two sides of the inequality in Step 5 we obtain $J(\mathsf{v}) - J(\mathsf{v}^0) = -\frac{5}{2}$ while $\propto w^0\mathsf{J}'(\mathsf{v}^0) \cdot \mathbf{s}^0 = -0.9$. Therefore we set $\mathsf{v}^1 := \mathsf{v}$ and return to Step 2. Since we had $J(\mathsf{v}^1) \neq 0$ we calculate $\mathbf{s}^1 = (-8, 0)^T$. Setting $w^1 = 1$ we obtain $\mathsf{v} := \mathsf{v}^1 + w^1\mathbf{s}^1 = (4, 0)^T$ which yields the new matrix game

$$H(\mathsf{v}, 1) = \begin{pmatrix} 6 & 7 \\ 5 & 4 \end{pmatrix}$$

which has value $T(\mathsf{v}, 1) = 6$ and basic optimal strategies $\mathsf{x}(\mathsf{v}, 1) = \mathsf{y}(\mathsf{v}, 1) = (1, 0)$. These yield $\mathsf{T}(\mathsf{v}) = (6, 0)^T, \Psi(\mathsf{v}) = (2, 0), J(\mathsf{v}) = 2$. Therefore the left side of the inequality in Step 5 is zero while the right side is negative. By Step 6 we now set $w^1 = \frac{1}{2}$, this leads to;

$$\mathsf{v} = \begin{pmatrix} 12 \\ 0 \end{pmatrix} + \frac{1}{2}\begin{pmatrix} 4 & 0 \\ 0 & 4 \end{pmatrix}\begin{pmatrix} -2 \\ 0 \end{pmatrix} = \begin{pmatrix} 8 \\ 0 \end{pmatrix}$$

and the new matrix game

$$H(\mathsf{v}, 1) = \begin{pmatrix} 9 & 8 \\ 8 & 7 \end{pmatrix},$$

$T(\mathsf{v}) = (8, 0)^T$ and $\Psi(\mathsf{v}) = \mathsf{T}(\mathsf{v}) - \mathsf{v} = 0, J(\mathsf{v}) = 0$. Thus $\mathsf{v}^2 := \mathsf{v}$ clearly satisfies the inequality in Step 5 and is indeed the value-vector of the stochastic game.

5. Conclusions and Open Problems: The demonstration that the Modified Newton's Method, that is, the MNM algorithm will solve an arbitrary discounted (or terminating) stochastic game opens the possibility of the numerical solution of these games on a moderate to large scale. Indeed, in preliminary numerical experiments we have solved some ten randomly generated problems with up to 180 states and up to 30 actions per player per state. Most of these were solved within accuracy of 10^{-5} in 4 (or less) of the major iterations.

Furthermore, our approach naturally suggests a new line of algorithmic investigations, based on the notion that increased efficiency could be gained by developing an appropriate (globally convergent?) Quasi-Newton algorithm (e.g., see

McCormick[4]) for solving discounted stochastic games. Similarly it is natural to ask to which classes (if any) of either zero-sum average, or general-sum discounted stochastic games could Newton-like algorithms be successfully applied. Finally, the implementation of the MNM algorithm, and of most related methods; is likely to raise interesting numerical analysis questions, the most obvious of which may be: What is the "best" way to avoid the computation of $(I - \beta P(\mathbf{v}))^{-1}$? Clearly, the answer to such a question may require not only the techniques of numerical analysis, but also insight into the structure of the stochastic game. For a related problem in the context of Markov Decision Processes an interesting approach to this problem was proposed by Puterman and Shin [6].

References

[1] *Breton, M.* (1987). Equilibre Pour Des Jeux Sequentiel, Ph.D. Thesis, University of Montreal.

[2] *Breton, M., J.A. Filar, A. Haurie, and T.A. Schultz* (1986). On the Computation of Equilibria in Discounted Stochastic Games, in Dynamic Games and Applications in Economics, T. Basar, editor, Springer-Verlag, Lecture Notes in Economics and Mathematical Systems.

[3] *Goldman, A.J. and A.W. Tucker* (1956). Theory of Linear Programming, in Linear Inequalities and Related Systems, H.W. Kuhn and A.W. Tucker, editors, Princeton University Press. Annals of Mathematics Studies #38.

[4] *McCormick, G.P.* (1983). Nonlinear Programming, Wiley, New York.

[5] *Pollatschek, M. and B. Avi-Itzhak* (1969). Algorithms for Stochastic Games with Geometric Interpretation, *Management Science*, **15**, 399-415.

[6] *Puterman, M.L. and M.C. Shin* (1978). Modified Policy Iteration Algorithms for Discounted Markov Decision Problems, *Management Science*, **24** , 1127-1137.

[7] *Rao, S.S and R. Chandrasekaran, and K.P.K. Nair* (1973). Algorithms for Discounted Stochastic Games, *Opt. Theory Appl.*, **11**, 627-637.

[8] *Shapley, L.S.* (1953). Stochastic Games, *Proceedings of the National Academy of Science U.S.A.*, **39**, 1095-1100.

[9] *Shapley, L.S. and R.N. Snow* (1950). Basic Solutions of Discrete Games, in Contributions to the Theory of Games, H.W. Kuhn and A.W. Tucker, editors, Princeton University Press, Annals of Mathematics Studies #24.

[10] *Van der Wal, J.* (1978). Discounted Markov Games: Generalized Policy Iteration Method, *J. Opt. Theory Appl.*, **25**, 125-138.

Jerzy A. Filar
Department of Mathematics
University of Maryland
Baltimore County
Baltimore, Maryland 21228
U.S.A.

Boleslaw Tolwinski
Department of Mathematics
Colorado School of Mines
Golden, Colorado 80401
U.S.A.

Semi-infinite Stochastic Games[1,2]

Sagnik Sinha, Frank Thuijsman and Stef H. Tijs

Abstract: We investigate two-person zero-sum stopping stochastic games with a finite number of states, for which the action sets of player I are finite and those for player II are countably infinite. Concerning the payoffs no restrictions are made. We show that for such games the value, possibly $-\infty$ in some coordinates, exists; player I possesses optimal stationary strategies and player II possesses near-optimal stationary strategies with finite support. Furthermore we relate the existence of value and of (near-)optimal stationary strategies with a maximal solution to the Shapley-equation.

1. Introduction: Let $M = [a_{ij}]_{i=1,j=1}^{m\ \ n}$ be an $m \times n$-matrix with $a_{ij} \in \mathbf{R}$ for all i and j. Then this matrix represents the following non-cooperative two-person game, called the matrix game M. There are two players, player I and player II, and a play in the game runs as follows. Simultaneously and independently of each other, player I chooses a row i and player II chooses a column j; then these choices are announced and player II has to pay the amount a_{ij} to player I.

Each of the players is allowed to randomize over pure action choices, which gives rise to the notion of mixed actions $p \in \Delta^m := \left\{x \in \mathbf{R}^m : x \geq 0, \sum_{i=1}^m x_i = 1\right\}$ for player I and mixed actions $q \in \Delta^n$ for player II. So, if player I uses the mixed action p and player II uses q, then with probability $p_i q_j$ entry (i, j) will be chosen, hence the expected payoff, by player II to player I, is pAq.

---

[1]This research started when Tijs was visiting the Indian Statistical Institute in New Delhi. We thank Professor T. Parthasarathy for his hospitality.

[2]Support by The Netherlands Organization for Scientific Research NWO, project 10-64-10, is gratefully acknowledged.

Now assuming that player I wants to maximize the expected payoff and player II wants to minimize the same, it is interesting to know if there exists something like the 'worth' of the matrix game M and 'best' mixed actions for the players. This problem was solved in 1928 by von Neumann [1] who showed that for each matrix M, as defined above, there exists a unique $v \in \mathbf{R}$ and further $p^* \in \Delta^m$ and $q^* \in \Delta^n$ such that for all $p \in \Delta^m$, $q \in \Delta^n$ it holds that $pAq^* \leq v \leq p^*Aq$. This v is called the value of the matrix game M and is denoted by $\mathrm{val}(M)$ or Let $M = \mathrm{val}[a_{ij}]_{i=1,j=1}^{m\ \ \ n}$ Further p^* is called an optimal mixed action for player I and q^* an optimal mixed action for player II. So using p^*, when repeating this matrix game M over and over again, guarantees player I's average income to be at least v; a similar interpretation can be made for player II.

In 1953 Shapley [2] introduced the idea of not playing one and the same matrix game every day, but playing one of a finite collection of matrix games every day (stage), with a motion along the matrix games from stage to stage governed by the current matrix game and the actions chosen there. To be more precise, there are z matrices $M^1, M^2, \ldots, M^z$, matrix M^s has size $m_s \times n_s$ and entry (i, j) of matrix M^s is given as:

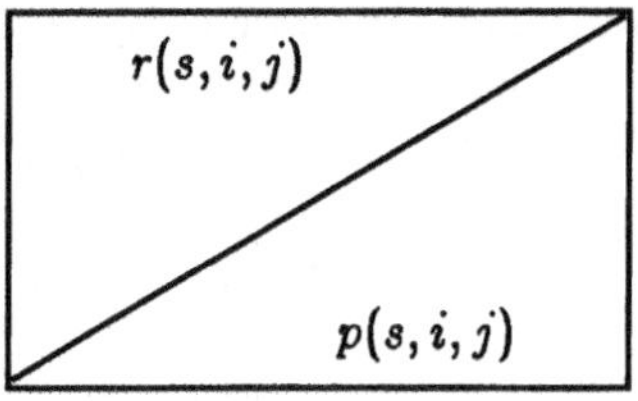

where $r(s,i,j) \in \mathbf{R}$ is a payoff to player I and

$$p(s,i,j) = (p(1|s,i,j), \ldots, p(z|s,i,j)) \in \Delta_o^z := \{x \in \mathbf{R}^z : x \geq 0, \sum_{s=1}^{z} x_s < 1\}.$$

The interpretation is: if in state s player I chooses $i \in A_s := \{1, 2, \ldots, m_s\}$ and player II chooses $j \in B_s := \{1, 2, \ldots, n_s\}$ then player II pays $r(s,i,j)$ to player I and next the play moves from state(matrix) s to state t with probability $p(t|s,i,j)$, and the play stops with probability $1 - \sum_{t=1}^{z} p(t|s,i,j) > 0$.

Thus, these 'stopping stochastic games' were naturally descended from matrix games. In stochastic games the players consider strategies; infinite plans that tell a player at each stage, at each state and for every history of the play, what mixed action to choose at that stage in that state. Strategies for which the prescribed mixed actions do not depend on the histories are called Markov strategies; Markov strategies for which the prescribed mixed actions do not depend on the stages are called stationary strategies. Stationary strategies are simply z-tuples of mixed actions, one mixed action for each state. It is obvious that stationary strategies are most easy to handle.

An initial state $s \in S$, together with a pair of strategies (π_1, π_2), for players I and II respectively, determine a stochastic process over the set of states $S = \{1, 2, \ldots, z\}$, and hence for all stages $\tau \in \mathbf{N}$ an expected direct payoff $R_{s\pi_1\pi_2}(\tau)$ by player II to player I is determined. The players evaluate this stream of expected payoffs $(R_{s\pi_1\pi_2}(1), R_{s\pi_1\pi_2}(2), \ldots)$ to be worth the reward $v(s, \pi_1, \pi_2) := \sum_{\tau=1}^{\infty} R_{s\pi_1\pi_2}(\tau)$.

By Shapley [2] it is known that there exists a unique $v \in \mathbf{R}^z$, a stationary strategy ρ^* for player I and a stationary strategy σ^* for player II such that for all π_1 and π_2 the following holds:

$$v(s, \pi_1, \sigma^*) \leq v(s) \leq v(s, \rho^*, \pi_2).$$

Here v is called the value of the stochastic game and $\rho^*(\sigma^*)$ is called an optimal stationary strategy for player I(II).

Furthermore, Shapley [2] shows that: v is the value of the stochastic game, ρ^* is an optimal stationary strategy for player I and σ^* an optimal stationary strategy for player II if and only if for all $s \in S$ one has that $v(s) = \mathrm{val}M_s(v)$ and that $\rho^*(s)$ is an optimal mixed action for player I in $M_s(v)$ and that $\sigma^*(s)$ is an optimal mixed action for player II in $M_s(v)$, where $M_s(v) = [r(s, i, j) + \sum_{t=1}^{z} p(t|s, i, j)v(t)]_{i=1,j=1}^{m_s\ n_s}$. Hence the value and optimal stationary strategies can be found by solving the set of equations: $x_s = \mathrm{val}M_s(x)$ for $s \in S, x \in \mathbf{R}^z$. This set of equations is known as the Shapley equation and can be shortened to $x = T(x)$ $(x \in \mathbf{R}^z)$, where T is the function from $\mathbf{R}^z$ to $\mathbf{R}^z$ defined by $(T(x))_s = \mathrm{val}M_s(x)$. The equation $x = Tx$ $(x \in \mathbf{R}^z)$ has a unique solution in $\mathbf{R}^z$ since T is a continuous contraction map.

Since 1953 the theory of stochastic games has been extended in many ways. Non-stopping stochastic games, i.e. stochastic games for which $p(s, i, j) \in \Delta^z$ for all s, i, j have been examined under the β-discounted as well as the limiting average criterion, two different ways of evaluating the infinite streams of expected payoffs; nonzero-sum stochastic games have been studied; stochastic games have been examined with infinite state and/or action spaces. This paper belongs to the last category. In this last category one mostly needs some boundedness conditions on the payoffs to be sure that the rewards are well-defined for all pairs of strategies. In this paper we investigate stopping stochastic games with a finite state space where for each state player I has a finite action space and player II has a countably infinite action space. We put no restrictions on the payoffs. Player II's strategy space however is restricted to the set of strategies having finite support, i.e. strategies for which there exists an $N \in \mathbf{N}$ such that for all states, stages and histories, the probability that player II will choose a column in $\{1, 2, \ldots, N\}$ equals 1. This is done to assure that all direct expected payoffs are well defined.

In section 2 we show that these semi-infinite stochastic games have a value in $\mathbf{R}_-^z$, where $\mathbf{R}_- = \mathbf{R} \cup \{-\infty\}$, that player I possesses optimal stationary strategies and that player II possesses near-optimal stationary strategies. In section 3 we relate solutions for these stochastic games with solutions of the Shapley equation in $\mathbf{R}_-^z$.

We finish this section with some examples to illustrate that these semi-infinite (stopping) stochastic games are the only reasonable countably infinite extension of the finite stochastic games as introduced in [2], if one wants to avoid putting some boundedness conditions on the payoffs.

Example 1.1: For a stochastic game with countable state space and finite action spaces one needs bounded payoffs to get well-defined rewards. Take for example the stochastic game with state space $\mathbf{N}$ and action set $\{1\}$ for both players in all states. The transition is given by $p(s+1|s,1,1) = \frac{1}{2}$ for all $s \in \mathbf{N}$ and $p(t|s,1,1) = 0$ otherwise; the payoffs to player I are given by $r(s,1,1) = (-2)^{s-1}$. Consider the game starting in state 1. In that case the reward $v(s,\pi_1,\pi_2)$ is not defined if one wants to take $\sum_{\tau=1}^{\infty} R_{s\pi_1\pi_2}(\tau)$.

Example 1.2: For a stochastic game with finite state space and both players having countably many actions in the same state, the value need not exist. Take for example the stochastic game with $S = \{1\}$, $A_1 = B_1 = \mathbf{N}$, the transition is given by $p(1|1,i,j) = \frac{1}{2}$ for all i, j, and the payoffs are given by $r(1,i,j) = 1$ if $i > j$, $r(1,i,j) = 0$ if $i = j$, $r(1,i,j) = -1$ if $i < j$. It is clear that $sup_{\pi_1} inf_{\pi_2} v(1,\pi_1,\pi_2) = -2$, whereas $inf_{\pi_2} sup_{\pi_1} v(1,\pi_1,\pi_2) = 2$.

Example 1.3: For a stochastic game with finite state space for which for each state one of the players has a finite action set, the value need not exist either. Consider the stochastic game with $s = \{1,2\}$; $A_1 = B_2 = \mathbf{N}$, $A_2 = B_1 = \{1\}$; $p(2|1,i,j) = \frac{1}{2} = p(1|2,i,j)$ for all i and j and $p(t|s,i,j) = 0$ else; $r(1,i,1) = i$, $r(2,1,j) = -j$ for all i and j.

If the players are not restricted to strategies with finite support, then $\sum_{\tau=1}^{\infty} R_{s\pi_1\pi_2}(R(\tau))$ need not exist. If the players are restricted to strategies with finite support, then

$sup_{\pi_1} inf_{\pi_2} v(s,\pi_1,\pi_2) = -\infty$ whereas $inf_{\pi_2} sup_{\pi_1} v(s,\pi_1,\pi_2) = +\infty$ for both $s = 1$ and $s = 2$.

2. Semi-infinite games: Tijs [3] examined semi-infinite matrix games and showed that for such games the value exists, though it may be $-\infty$ if player II is the player with action set **N**. For player II one cannot allow all probability distributions over **N** as mixed actions without encountering difficulties concerning the expected direct payoffs. Tijs [3] considers several restrictions on player II's set of mixed actions. One of them is restricting player II to use only mixed actions with finite support, i.e. mixed actions in $\bigcup_{n\in\mathbf{N}} \Delta^n$; where we identify Δ^n with the set $\{x \in \mathbf{R}^\infty : x \geq 0, \sum_{i=1}^n x_i = 1,\ x_i = 0 \text{ for } i > n\}$.

In this section we extend the work of Tijs [3] on semi-infinite matrix games, with the above restriction for player II, to semi-infinite stochastic games.

Preliminaries: A semi-infinite stochastic game Γ_∞ is given by a finite set of states S, where for each $s \in S$ there is a $m_s \times \infty$ matrix M^s of which entry (i, j) contains $r(s,i,j) \in \mathbf{R}$, the payoff to player I, and $p(s,i,j) \in \Delta_o^z$, a transition vector. However, we assume that $sup_{s,i,j}\{\sum_{t\in S} p(t|s,i,j)\} < 1$. Play proceeds in stages as explained in section 1. Player II is restricted to strategies π_2 with finite support, i.e. for each π_2 there exists an $N \in \mathbf{N}$ such that, before stopping, with probability 1, player II will choose one of the first N columns at all stages, in all states, for all histories.

Related with Γ_∞, for all $n \in \mathbf{N}$ one can look at the n-truncated stochastic game Γ_n, which we get by deleting all columns j with $j > n$, for all states. By completion with zeros, strategies for player II in Γ_n can be identified with finite support strategies for player II in Γ_∞; similarly, by deleting zeros, finite support strategies for player II in Γ_∞ can be identified with strategies for player II in Γ_n for n sufficiently large. We also identify Markov strategies for player I in Γ_∞ with those in Γ_n, $n \in \mathbf{N}$.

If there exists a $v_\infty \in \mathbf{R}_-^z$ such that for all $\epsilon > 0$ and all $\delta < 0$ there are strategies $\pi_{\epsilon\delta}^1$ and $\pi_{\epsilon\delta}^2$ for the respective players for which for all $s \in S$, all π^1, π^2 (with finite support)

$$v_\infty(s, \pi_{\epsilon\delta}^1, \pi^2) \geq v_\infty(s) - \epsilon$$

$$v_\infty(s, \pi^1, \pi_{\epsilon\delta}^2) \leq \begin{cases} v_\infty(s) + \epsilon & \text{if } v_\infty(s) \in \mathbf{R} \\ \delta & \text{if } v_\infty(s) = -\infty \end{cases}$$

then v_∞ is called the value of Γ_∞, and also $\pi_{\epsilon\delta}^1$ is called an $\epsilon\delta$-optimal strategy for player I and similarly $\pi_{\epsilon\delta}^2$is called an $\epsilon\delta-$ optimal strategy for player II. If one can take $\epsilon = 0$ and $\delta = -\infty$, then such $\epsilon\delta$-optimal strategies are called optimal strategies.

As we know the value v_n of Γ_n exists for all n.

Theorem 2.1:

(i) $lim_{n\to\infty} v_n$ exists in $\mathbf{R}_-^z$ and equals v_∞, the value of Γ_∞.

(ii) Player I has optimal stationary strategies in Γ_∞.

(iii) Player II has near-optimal stationary strategies in Γ_∞.

Proof: For all $n \in \mathbf{N}$ it holds that $v_n \geq v_{n+1}$, coordinatewise, since in Γ_{n+1} player II's action set is larger than in Γ_n, and player I's action set remains the same. As $v_1, v_2, \ldots$ is non-increasing it converges to some $w \in \mathbf{R}^z_-$, coordinatewise. We prove that w is the value of Γ_∞.

For every $n \in \mathbf{N}$ player I has an optimal stationary strategy ρ^n in Γ_n. All ρ^n, $n \in \mathbf{N}$, can be seen as elements of the compact set $\Delta^{m_1} \times \Delta^{m_2} \times \ldots \times \Delta^{m_z}$. Hence, without loss of generality we can assume that $\rho^1, \rho^2, \ldots$ converges to some stationary strategy ρ^*. Now let π^2 be any strategy with finite support for player II. Then π^2 can be seen as a strategy for player II in Γ_n for n sufficiently large. Fix n_0 large enough. It is well-known that playing against a fixed stationary strategy in a stopping stochastic game with finite state and action spaces, Γ_{n_0} for instance, an optimal reply can be found among the stationary strategies. Blackwell [4] proves this for the special class of β-discounted stochastic games, but his proof can be applied to stopping games as well. Hence there is a σ for player II in Γ_{n_0} such that $v(\rho^*, \sigma) \leq v(\rho^*, \pi^2)$, coordinatewise. We also know $v(\rho^n, \sigma) \geq v_n$ for all $n \geq n_0$. Now, using the fact that $\rho \mapsto v(\rho, \sigma)$ is a continuous function on the set of stationary strategies of player I, and using the fact that $v_1, v_2, \ldots$ converges to w, we conclude $v(\rho^*, \sigma) \geq w$, and finally $v(\rho^*, \pi^2) \geq w$. So the lower value of Γ_∞ is at least w.

Take $\epsilon > 0$, $\delta < 0$. Since $v_1, v_2, \ldots$ converges to w there exists $n_1 \in \mathbf{N}$ such that $v_{n_1}(s) \leq w(s) + \epsilon$ if $w(s)$ is finite, and $v_{n_1}(s) \leq \delta$ if $w(s) = -\infty$, for all $s \in S$. Let σ^* be an optimal stationary strategy for player II in Γ_{n_1}. Then again applying the argument that against σ^* player I has a stationary strategy as best answer ([4]), we have that for all strategies π^1 for player I and all $s \in S$ it holds that: $v(s, \pi^1, \sigma^*) \leq v_{n_1}(s) \leq w(s) + \epsilon$ if $w(s)$ is finite; $v(s, \pi^1, \sigma^*) \leq v_{n_1}(s) \leq \delta$ if $w(s) = -\infty$. So the upper value of Γ_∞ is at most w. Combination of the above arguments proves the theorem. ◇

It is clear that player II needs not possess an optimal stationary strategy (with finite support). A simple example for this is the stochastic game consisting of one $1 \times \infty$ state matrix for which $r(1, 1, j)$ equals $-j$ and $p(1, 1, j) = 0$ for all $j \in \mathbf{N}$.

In the following theorem we give a necessary and sufficient condition for player II to possess optimal stationary strategies. First we introduce the concept 'critical number' for a semi-infinite stochastic game Γ_∞. The critical number c is defined by $c = min\{n \in \mathbf{N} : v_n = v_\infty\}$, where $min\, \emptyset = \infty$. If c is finite, then Γ_∞ is called an essentially finite game.

Theorem 2.2: A semi-infinite stochastic game Γ_∞ is essentially finite if and only if player II has optimal stationary strategies.

Proof: Suppose $c < \infty$. Let σ^* be an optimal stationary strategy for player II in Γ_c. Playing against σ^* in Γ_∞ player I has a stationary best answer ([4]). For all stationary ρ we have $v(\rho, \sigma^*) \leq v_c = v_\infty$, so for all strategies π^1 for player I in $\Gamma_\infty : v(\pi^1, \sigma^*) \leq v_\infty$. Hence σ^* is an optimal stationary strategy for player II in Γ_∞.

Starting with a stationary optimal strategy $\hat{\sigma}$ for player II in Γ_∞ which only uses the first n_1 columns, it is clear that for all stationary strategies ρ for player I we have $v(\rho, \hat{\sigma}) \leq v_\infty$. Hence $v_{n_1} \leq v_\infty$. But for all $n \in \mathbf{N}$ we also had $v_n \geq v_\infty$. So $v_{n_1} = v_\infty$ and therefore $c \leq n_1 < \infty$. ◇

For essentially finite stochastic games Γ_∞ we look at the relation between sets of optimal stationary strategies in Γ_∞, O^1_∞ and O^2_∞, and the sets of optimal stationary strategies in Γ_n, O^1_n and O^2_n, $n \in \mathbf{N}$. Note that for an essentially finite stochastic game both O^1_n and O^2_n are non-empty for all $n \in \mathbf{N} \cup \{\infty\}$ by Theorems 2.1 and 2.2.

Theorem 2.3: Let Γ_∞ be an essentially finite stochastic game. Then $O^1_\infty = \bigcap_{n \geq c} O^1_n$ and $O^2_\infty = \bigcup_{n \geq c} O^2_n$.

Proof: By the proof of Theorem 2.1 we know that, if some sequence of stationary strategies $\rho^1, \rho^2, \ldots$, with $\rho^n \in O^1_n$ for all $n \in \mathbf{N}$, converges to a stationary strategy ρ^*, then $\rho^* \in O^1_\infty$. So we have $O^1_\infty \supset \bigcap_{n \geq c} O^1_n$. Conversely, if $\rho^* \in O^1_\infty$ then for all stationary strategies σ for player II in Γ_n, $n \geq c$, it holds that $v(\rho^*, \sigma) \geq v_\infty = v_n$. Hence $O^1_\infty \subset \bigcap_{n \geq c} O^1_n$. Combining we have proved the player I part of the theorem.

The player II part follows directly from the proof of Theorem 2.2. ◇

3. The Shapley Equation: As already mentioned in the introduction, for finite two-person stopping stochastic games the value and optimal stationary strategies can be found by solving the Shapley equation $x = Tx$ $(x \in \mathbf{R}^z)$, where

$$(Tx)_s = \text{val} M_s(x) = \text{val}[r(s,i,j) + \sum_{t=1}^{z} p(t|s,i,j) x_t]_{i=1,j=1}^{m_s, n_s}$$

which, in the finite case, has a unique solution.

One can also consider the Shapley equation for semi-infinite stochastic games. The difference is that for these stochastic games the $M_s(x)$ will be semi-infinite matrices and the system $x = Tx$ should be solved over $\mathbf{R}^z_-$. Examining $x = Tx$ $(x \in \mathbf{R}^z_-)$ one quickly notes that $(-\infty, -\infty, \ldots, -\infty)$ will often be a solution, hence unlike in the finite case we may no longer have a unique solution. However, some interesting results can be derived.

Theorem 3.1: The value of the semi-infinite stochastic game, v_∞, is a solution of the Shapley equation.

Proof: For all $x \in \mathbf{R}_-^z$ and all $s \in S$ let $M_s(x)$ denote the semi-infinite matrix game with entries $(r(s,i,j) + \sum_{t=1}^{z} p(t|s,i,j)x_t)$ in $\mathbf{R}_-$, and let $M_s^n(x)$ denote the corresponding n-truncated matrix game for $n \in \mathbf{N}$. We know by the result of Shapley [2] that $v_n(s) = \text{val}M_s^n(v_n)$ for all $s \in S$.

First, suppose $v_\infty \in \mathbf{R}^z$ and take $\epsilon > 0$. There exists $n_0 \in \mathbf{N}$ such that for all $n \geq n_0$:

(i) $|v_\infty(s) - v_n(s)| < \epsilon$ for all $s \in S$, because $lim_{n\to\infty} v_n(s) = v_\infty(s) \in \mathbf{R}$ for all $s \in S$

(ii) $|\text{val}M_s^n(v_n) - \text{val}M_s^n(v_\infty)| < \epsilon$ for all $s \in S$, because payoffs in $M_s^n(v_n)$ differ less than ϵ from corresponding payoffs in $M_s^n(v_\infty)$ for all $s \in S$.

(iii) $|\text{val}M_s^n(v_\infty) - \text{val}M_s(v_\infty)| < \epsilon$ for all $s \in S$, because $lim_{k\to\infty} \text{val}M_s^k(v_\infty) = \text{val}M_s(v_\infty)$.

Combining (i), (ii) and (iii) yields $|v_\infty(s) - \text{val}M_s(v_\infty)| < 3\epsilon$ for all $s \in S$. Since ϵ was arbitrary we have $v_\infty(s) = \text{val}M_s(v_\infty)$ for all $s \in S$.

Second, if $v_\infty \notin \mathbf{R}^z$ then, without loss of generality, there is $k \in \{0, 1, \ldots, z-1\}$ such that $v_\infty(s) = -\infty$ if and only if $s \in \{k+1, k+2, \ldots, z\}$. For all $s \in \{1, 2, \ldots, k\}$ player I can prevent the play to move to any state outside $\{1, 2, \ldots, k\}$, otherwise $v_\infty(s)$ would be $-\infty$ as well. So for $s \in \{1, 2, \ldots, k\}$ player I can restrict to rows i in $M_s(v_\infty)$ for which $p(t|s,i,j) = 0$ for all $t \in \{k+1, k+2, \ldots, z\}$ and all $j \in \mathbf{N}$. Hence, restricting player I's actions this way yields a stochastic game $\tilde{\Gamma}$ for which play remains in $\{1, 2, \ldots, k\}$ forever. It is obvious that the value of this new stochastic game with state space $\{1, 2, \ldots, k\}$ equals the value of the original stochastic game on those states. So the value of the new stochastic game is finite and as we have just seen we can conclude $v_\infty(s) = \text{val}M_s(v_\infty)$ for $s \in \{1, 2, \ldots, k\}$. If $v_\infty(s) = -\infty$ for some $s \in S$, then this implies that player I cannot force a play to remain within the set of states $\{1, 2, \ldots, k\}$ with probability 1. From this one can conclude that in each row of $M_s(v_\infty)$ some entries equal $-\infty$. Hence $v_\infty(s) = \text{val}M_s(v_\infty)$ for all states (with value $-\infty$ or otherwise). ◇

Theorem 3.2: If $x \in \mathbf{R}^z$ is a solution to the Shapley equation, then $x = v_\infty$.

Proof: Suppose $x \in \mathbf{R}^z$ is a solution to the Shapley equation. Then for all $s \in S : x_s = \text{val}M_s(x) = \text{val}[r(s,i,j) + \sum_{t=1}^{z} p(t|s,i,j)x_t]_{i=1,j=1}^{m_s,\infty}$.

Let ρ_s^* be an optimal mixed action for player I in $M_s(x)$ for all $s \in S$. Then for all stationary strategies σ for player II it holds that $x_s \leq r(s,\rho_s^*,\sigma_s) + \sum_{t=1}^{z} p(t|s,\rho_s^*,\sigma_s)x_t$, where $r(s,\rho_s^*,\sigma_s) = \sum_{i=1}^{m_s}\sum_{j=1}^{\infty} \rho_s^*(i)r(s,i,j)\sigma_s(j)$ and $p(t|s,\rho_s^*,\sigma_s) = \sum_{i=1}^{m_s}\sum_{j=1}^{\infty} \rho_s^*(i)p(t|s,i,j)\sigma_s(j)$. In vector-notation, letting $r(\rho^*,\sigma) = (r(1,\rho_1^*,\sigma_1), \ldots r(z,\rho_z^*,\sigma_z))$ and $P(\rho^*,\sigma)$ the $z \times z$-matrix with $p(t|s,\rho_s^*,\sigma_s)$ in entry (s,t), we get $x \leq r(\rho^*,\sigma) + P(\rho^*,\sigma)x$. This implies $x \leq \sum_{\tau=1}^{n} P^{\tau-1}(\rho^*,\sigma)r(\rho^*,\sigma) + P^n(\rho^*,\sigma)x$ for all $n \in \mathbf{N}$, where $P^\tau(\rho^*,\sigma)$ is the τ-fold product of $P(\rho^*, \sigma)$ and $P^0(\rho^*, \sigma)$ is the identity matrix. Since by the stopping play assumption, $lim_{n\to\infty} P^n(\rho^*, \sigma)x = 0$ we have $x \leq \sum_{\tau=1}^{\infty} P^{\tau-1}(\rho^*, \sigma)r(\rho^*, \sigma) = v(\rho^*, \sigma)$. Hence $x \leq v_\infty$.

Conversely, let σ_s^* be an ϵ-optimal mixed action for player II in $M_s(x)$, for all $s \in S$. Then we have that for all stationary strategies ρ for player I: $x \geq r(\rho, \sigma^*) + P(\rho, \sigma^*)x - \epsilon 1_z$. From this one can derive $x \geq v(\rho, \sigma^*) - \epsilon(1-\alpha)^{-1}1_z$ where $\alpha = max_{s,i} \sum_{t=1}^{z} p(t|s, i, \sigma_s^*) < 1$. Since this can be done for all $\epsilon > 0$, we have $x \geq v_\infty$. ◇

Observe that Theorem 3.2 says, that there is at most one real solution to the Shapley equation, and if there is one, it necessarily equals v_∞.

The following theorem tells us how to find v_∞ among all solutions.

Theorem 3.3: If $x \in \mathbf{R}_-^z$ is a solution to the Shapley equation, then $x \leq v_\infty$.

Proof: Let $x \in \mathbf{R}_-^z$ be a solution. For $s \in S$ with $x_s = -\infty$, it is clear that $x_s \leq v_\infty(s)$. For $s \in S$ with $x_s \in \mathbf{R}$ we have $x_s = \text{val}\left[r(s, i, j) + \sum_{t=1}^{z} p(t|s, i, j)x_t\right]_{i=1, j=1}^{m_s\ \infty}$. This implies that for an optimal mixed action ρ_s^* of player I in $M_s(x)$ we have $\sum_{t=1}^{z} p(t|s, \rho_s^*, j)x_t \in \mathbf{R}$ for all $j \in \mathbf{N}$, and hence $p(t|s, \rho_s^*, j) = 0$ for all t with $x_t = -\infty$ and all $j \in \mathbf{N}$. Without loss of generality let $k \in S$ be such that $x_t = -\infty$ for $t > k$ and $x_t \in \mathbf{R}$ for $t \leq k$. Then, for $s \in \tilde{S} := \{1, 2, \dots, k\}$, we have that for all stationary strategies σ for player II: $x_s \leq r(s, \rho_s^*, \sigma_s) + \sum_{t=1}^{k} p(t|s, \rho_s^*, \sigma_s)x_t$. In vector notation over $\tilde{S} : x \leq r(\rho^*, \sigma) + P(\rho^*, \sigma)x$. Now, using $x_s \in \mathbf{R}$ for all $s \in \tilde{S}$, iteration of the above inequality gives $x \leq v(\rho^*, \sigma)$. Hence we have shown that $x_s \leq v_\infty(s)$ for all $s \in \tilde{S}$. ◇

Theorem 3.4: A stationary strategy ρ^* for player I is optimal if and only if for each $s \in S$ the mixed action ρ_s^* is optimal in the semi-infinite matrix game $M_s(v_\infty)$.

Proof: If for each $s \in S$ the mixed action ρ_s^* is optimal in $M_s(v_\infty)$, then, by the proof of Theorem 3.3, it follows that ρ^* is an optimal stationary strategy in the stochastic game.

Conversely, let ρ^* be an optimal stationary strategy for player I. Suppose, for some $s \in S$, that ρ_s^* is not optimal in $M_s(v_\infty)$, then, for some $j \in \mathbf{N}$ and some $\epsilon > 0$, it holds that $r(s, \rho_s^*, j) + \sum_{t=1}^{z} p(t|s, \rho_s^*, j)v_\infty(t) < \text{val}M_s(v_\infty) - \epsilon = v_\infty(s) - \epsilon$. Hence, if for the stochastic game starting in s, player II initially chooses column j and from then on uses an $\epsilon\delta$-optimal stationary strategy $\pi_{\epsilon\delta}^2$ against ρ^*, then for this strategy $(j, \pi_{\epsilon\delta}^2)$ of player II we have $v(s, \rho^*, (j, \pi_{\epsilon\delta}^2)) \leq r(s, \rho_s^*, j) + \sum_{t=1}^{z} p(t|s, \rho_s^*, j)(v_\infty(t) + \epsilon) < v_\infty(s)$. This contradicts the optimality of ρ^* in the stochastic game. ◇

For player II near-optimal stationary strategies cannot directly be found as extensions of mixed actions in the matrix games $M_s(v_\infty)$. Think for instance of the stochastic game consisting of one $1 \times \infty$ state with $r(1,1,j) = -j$ and $p(1|1,1,j) = \frac{1}{2}$ for all $j \in \mathbf{N}$. It is obvious that $v_\infty = -\infty$. Hence $M_1(v_\infty)$ is the $1 \times \infty$ matrix game with payoff $-\infty$ in all entries. Clearly, the mixed action: "choose column j" is an optimal mixed action for player II in the matrix game $M_1(v_\infty)$, but does not give any information about near-optimal strategies in the stochastic game. This is due to the fact that a state s, with $v_\infty(s) = -\infty$, is either 'directly good' or 'indirectly good' for player II. The following two lemmas illustrate this phenomenon.

Lemma 3.5: If $v_\infty \notin \mathbf{R}^z$ and $\delta < 0$, then there is at least one state $s \in S$ in which player II has a mixed action q_s, such that the expected direct payoff in state s is at most δ, if player II uses the mixed action q_s.

Proof: If not, then player I has a stationary strategy ρ such that all expected direct payoffs are at least δ, and hence $v \in \mathbf{R}^z$. Contradiction. ◇

Let $D \subset S$, be the set of states for which player II has, for all $\delta < 0$, a mixed action to keep the expected direct payoff below δ. If $v_\infty \notin \mathbf{R}^z$, then D is non-empty by Lemma 3.5. D is called the set of states that are directly good for player II.

Lemma 3.6: If $v_\infty \notin \mathbf{R}^z$, then let $ID = \{s \in S \backslash D : v_\infty(s) = -\infty\}$ and suppose $ID \neq \emptyset$. Then player II has mixed actions q_s, $s \in ID$, such that for all stationary strategies σ for player II with $\sigma_s = q_s$ for $s \in ID$, and all stationary strategies for player I, any play started in some state in ID will reach the set of states D with positive probability.

Proof: If not, then $v_\infty(s) \in \mathbf{R}$ for some $s \in ID$. Contradiction. ◇

ID may be empty and is called the set of indirectly good states for player II. Player II can construct a near-optimal strategy in the following way. In directly good states (belonging to D), player II ensures that the expected directed payoff is low enough; in indirectly good states (belonging to ID) player II ensures that the transitions will lead to D; in the other states, with finite value, player II has to consider direct payoffs as well as transitions.

Theorem 3.7: A near-optimal stationary strategy for player II can be constructed by taking mixed actions q_s, $s \in S$, which are near-optimal in the matrix game $[r(s,i,j)]_{i=1,j=1}^{m_s,\infty}$ for $s \in D$, in the matrix game $[r(s,i,j) + \sum_{t=1}^{z} p(t|s,i,j)v_\infty(t)]_{i=1,j=1}^{m_s,\infty}$ for $s \in S\backslash(D \cup ID)$, and for $s \in ID(k)$ to be taken near-optimal in the matrix game $\left[\sum_{t=1}^{z} p(t|s,i,j)w^k(t)\right]_{i=1,j=1}^{m_s,\infty}$, where $ID(k) = \{s \in ID$; player I cannot avoid the play to move from s to D within k stages, with positive probability}, and where $w^k(t) = -\infty$ for $t \in D \cup \left(\bigcup_{\ell=1}^{k} ID(\ell)\right)$ and $w^k(t) = 0$ for other $t, k \in \{0, 1, 2, \ldots, z-1\}$.

Proof: Follows straightforward from Lemma 3.5, Lemma 3.6 and part of the proof of Theorem 3.2. ◇

For stochastic games with finite state and action spaces and any $x \in \mathbf{R}^z$ it holds that $lim_{n\to\infty} T^n(x) = v_\infty$, by the facts that T is a contraction operator and v_∞ is its unique fixed point. Since for semi-infinite stochastic games the value may be $-\infty$ in some coordinates, it is not directly clear whether the above method can be used to find the value. The following theorem answers this problem.

Theorem 3.8: For all $x \in \mathbf{R}^z$, $lim_{n\to\infty} T^n(x) = v_\infty$.

Proof: $D = \{s \in S : (T(x))_s = -\infty\}$, $ID = \{s \in S \backslash D : (T^z(x))_s = -\infty\}$. If $S = D \cup ID$, we are finished, else we continue in the following way. For the remaining states $(T^z(x))_S \in \mathbf{R}$ and hence $(T^n(x))_s \in \mathbf{R}$ for all these remaining states. Let $R := S \backslash (D \cup ID)$. Without loss of generality suppose $R = \{1, 2, \ldots, k\}$. For $y = (y_1, y_2, \ldots, y_k, -\infty, \ldots, -\infty) \in \mathbf{R}_-^z$, with $\hat{y} = (y_1, y_2, \ldots, y_k) \in \mathbf{R}^k$, and $s \in R$ it holds that $(T(y))_s = (\widehat{T}(\hat{y}))_s$ where

$$(\widehat{T}(\hat{y}))_s = \text{val}\left[r(s,i,j) + \sum_{t=1}^{k} p(t|s,i,j)\hat{y}_t\right]_{i=1,j=1}^{m_s(R)\infty} =: \text{val}\widehat{M}_s(\hat{y}),$$

where, without loss of generality, $\{1, 2, \ldots, m_s(R)\}$ is the set of rows of $M_s(y)$ for which all entries are reals, and where $\widehat{M}_s(\hat{y})$ is the matrix consisting of those rows.

For $a, b \in \mathbf{R}^k$, $a \neq b$, it holds that $\|\widehat{T}(a) - \widehat{T}(b)\| < \|a - b\|$, where $\|x\| = max_s |x_s|$. Hence $\widehat{T}$ is a contraction map on $\mathbf{R}^k$ and has a unique fixed point which necessarily equals $lim_{n\to\infty} \widehat{T}^n(\hat{x})$ for all $\hat{x} \in \mathbf{R}^k$. For $s \in R$, player I's optimal stationary strategies never lead the play to $D \cup ID$ and hence, starting in R, the stochastic game can be seen as a stochastic game with R as set of states: $\widehat{\Gamma}$. Simply restrict player I's action sets as is done above. Applying Theorems 3.1 and 3.2 to $\widehat{\Gamma}$ we have that $v_\infty(s) = lim_{n\to\infty}(\widehat{T}^n(\hat{x}))_s = lim_{n\to\infty}(T^n(x))_s$ for $s \in R$. For $s \in D \cup ID$ and $n \geq z : (T^n(x))_s = -\infty = v_\infty(s)$. Hence $lim_{n\to\infty} T^n(x) = v_\infty$. ◇

Observe that $lim_{n\to\infty} T^n(x)$ need not equal v_∞ if we start with $x \in \mathbf{R}_-^z \backslash \mathbf{R}^z$. This is illustrated by the following example:

Take $S = \{1,2,3\}$; $A_1 = \{1,2\}$, $A_2 = A_3 = \{1\}$, $B_1 = B_2 = B_3 = \mathbf{N}$; $r(1,1,j) = r(2,1,j) = r(3,1,j) = 0$ for all $j \in \mathbf{N}$, $r(1,2,j) = 2$ for all $j \in \mathbf{N}$; $p(1|1,1,j) = p(3|1,2,j) = p(1|2,1,j) = p(3|3,1,j) = \frac{1}{2}$, $p(t|s,i,j) = 0$ else. Then, starting with $x = (0,-\infty,-\infty)$, it follows that $T^n(x) = (0,0,-\infty)$ for all $n \in \mathbf{N}$, where as $v = (2,1,0)$.

Closing Remarks: In the literature β-discounted stochastic games have been studied extensively. Those are stochastic games that continue infinitely, since there one takes in the definition of the game that $\sum_{t=1}^{z} p(t|s,i,j) = 1$ for all s,i,j. The players however, discount future payoffs by some factor $\beta \in (0,1)$. So for a pair of strategies (π_1,π_2) and an initial state s the β-discounted reward to player I is given by $\sum_{\tau=1}^{\infty} \beta^{\tau-1} R_{s\pi_1\pi_2}(\tau)$.

Notice that a β-discounted stochastic game can be seen as a stopping stochastic game if we relate transition probabilities $p(t|s,i,j)$ in the β-discounted non-stopping game with transition probabilities $\beta p(t|s,i,j)$ in the stopping game.

Hence all the results derived in this paper for stopping stochastic games, also hold for β-discounted stochastic games.

In section 2 we made the assumption $sup_{s,i,j}\{\sum_{t\in S} p(t|s,i,j)\} < 1$. This condition played no role until Lemma 3.5.

References

[1] *Von Neumann, J.* (1928). Zur Theorie der Gesellschaftsspiele, *Mathematische Annalen*, **100**, 295-320.

[2] *Shapley, L.S.* (1953). Stochastic Games. *Proceedings of the National Academy of Sciences, U.S.A.*, **39**, 1095-1100.

[3] *Tijs, S.H.* (1975). Semi-Infinite and Infinite Matrix Games and Bimatrix Games. Ph.D. Thesis, University of Nijmegen.

[4] *Blackwell, D.* (1962). Discrete Dynamic Programming. *Annals of Mathematical Statistics*, **33**, 719-726.

S. Sinha
Indian Statistical Institute
7 S.J. Sansanwal Marg
New Delhi 110016
India.

F. Thuijsman
Department of Mathematics
Faculty of General Sciences
University of Limburg
P.O. Box 616
6200 MD Maastricht
The Netherlands.

S.H.Tijs
Department of Mathematics
Catholic University
Toernooiveld,
6525 ED Nijmegen
The Netherlands

Easy Initial States in Stochastic Games[1]

Frank Thuijsman and Okko J. Vrieze

Abstract: In this paper we deal with limiting average stochastic games with finite state and action spaces. For any nonzero-sum stochastic game of this type, there exists a subset of initial states for which an almost stationary ϵ-equilibrium exists. For any zero-sum stochastic game there exists for each player a subset of initial states for which this player has an optimal stationary strategy.

1. The Stochastic Game Model; Definitions and Notations: In this paper we consider stochastic games with finite state and action spaces. For sake of notational convenience we focus on two-person stochastic games, though our approach can also be applied to N-person stochastic games. The players are called player I and player II.

A stochastic game situation can be seen as a finite set of matrices $\{L_1, L_2, \ldots, L_z\}$, corresponding to the set of states $S := \{1, 2, \ldots, z\}$, in which entry (i, j) of L_s, $s \in S$, has the following shape:

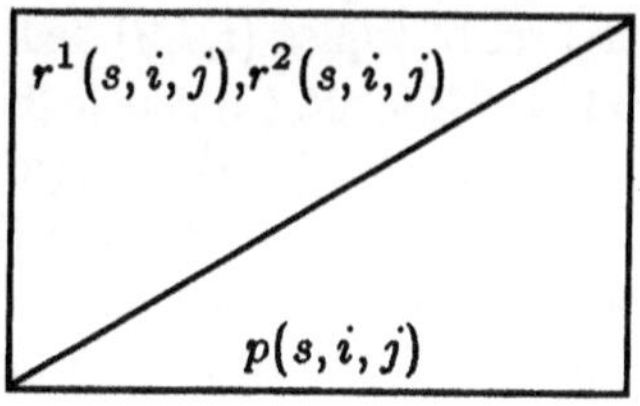

[1]Support was provided by the Netherlands Foundation for Mathematics SMC with financial support by the Netherlands Organization for Scientific Research NWO (project 10-64-10).

For all s, i and j : $r^1(s,i,j), r^2(s,i,j) \in \mathbf{R}$ and $p(s,i,j) := (p(1|s,i,j), p(2|s,i,j), \ldots, p(z|s,i,j))$ is a probability vector in $\mathbf{R}^z$.

The stochastic game corresponding to this situation is played along discrete stages, denoted by $\mathbf{N} := \{1, 2, \ldots\}$, in the following way. At each stage $n \in \mathbf{N}$ the system is in one of the states. Say it is in state s at stage n. Simultaneously and independently player I chooses a row of L_s, i for instance, and player II chooses a column of L_s, j for instance. These choices may be seen as outcomes of probability experiments done by the players. Then the players are informed of each other's choices, player I receives the immediate payoff $r^1(s,i,j)$, player II receives the immediate payoff $r^2(s,i,j)$ and next the system moves to a subsequent state determined by a probability experiment according to $p(s,i,j)$, i.e. a move to state t occurs with probability $p(t|s,i,j)$, for each $t \in S$. In this new state, announced to the players, choices have to be made by the players at stage $n+1$, etc.

A player's strategy is determined by a specification of probability distributions at each stage over his available actions, conditional on the history of the game up to that stage. Note that we assume the players to have complete information about states and actions that occurred in the past. A strategy for player I will be denoted by π_1, for player II by π_2. A strategy is a Markov strategy if the specified probability distribution at each stage only depends on the current state and on the stage number. A strategy is called a stationary strategy if it is a Markov strategy and if the specified probability distributions are the same for all stages. Stationary strategies are denoted by τ for player I and by σ for player II.

Observe that stationary strategies can be considered as z-tuples, $\tau = (\tau_1, \tau_2, \ldots, \tau_z)$ and $\sigma = (\sigma_1, \sigma_2, \ldots, \sigma_z)$ respectively, where τ_s and σ_s, $s \in S$, are probability distributions over the sets of available actions for player I and player II respectively. The action sets are denoted by $A_s := \{1, 2, \ldots, m_s\}$ and $B_s := \{1, 2, \ldots, n_s\}$ in state $s \in S$ for player I and player II respectively. Then $\tau_s = (\tau_s(1), \tau_s(2), \ldots, \tau_s(m_s))$ with $\tau_s(i) \geq 0$ and $\sum_{i=1}^{m_s} \tau_s(i) = 1$; and $\sigma_s = (\sigma_s(1), \sigma_s(2), \ldots, \sigma_s(n_s))$ with $\sigma_s(j) \geq 0$ and $\sum_{j=1}^{n_s} \sigma_s(j) = 1$.

A pair of strategies (π_1, π_2) together with an initial state $s \in S$ determine a stochastic process on the set of triples (t, i, j) and thus a stochastic process on the payoffs $r^k(t,i,j)$. For all $n \in \mathbf{N}$ let $R_n(k)$ denote the stochastic variable representing the immediate payoff to player $k \in \{1, 2\}$ at stage n. Expectations with respect to s, π_1, π_2 will be denoted by $E_{s\pi_1\pi_2}$. Since we will assume that both players want to maximize their overall reward, the players have to use some criterion in order to compare streams of payoffs. The two main evaluation criteria are β-discounted, where players discount future payoffs, and limiting average, where players, roughly speaking, evaluate by looking at the long run average reward. To be more exact: for a pair of strategies (π_1, π_2) and initial state s

$$v_\beta^k(s, \pi_1, \pi_2) = E_{s\pi_1\pi_2}\left(\sum_{n=1}^{\infty} \beta^{n-1} R_n(k)\right)$$

is the β-discounted reward to player k, $\beta \in [0,1)$;

$$g^k(s,\pi_1,\pi_2) = E_{s\pi_1\pi_2}\left(\liminf_{N\to\infty} \frac{1}{N}\sum_{n=1}^{N} R_n(k)\right)$$

is the limiting average reward to player k.

Depending on which criterion is used (both players use the same) the game is called a β-discounted stochastic game or a limiting average stochastic game.

For any pair of strategies it holds that $E_{s\pi_1\pi_2}(\sum_{n=1}^{\infty}\beta^{n-1}R_n(k))$ is equal to $\sum_{n=1}^{\infty}\beta^{n-1}E_{s\pi_1\pi_2}(R_n(k))$, whereas $E_{s\pi_1\pi_2}(\liminf_{N\to\infty}\frac{1}{N}\sum_{n=1}^{N}R_n(k))$ does not necessarily equal $\liminf_{N\to\infty}\frac{1}{N}\sum_{n=1}^{N}E_{s\pi_1\pi_2}(R_n(k))$. However, for stationary strategies the latter two expressions are equal, and furthermore in that case, "lim inf" can be replaced by "lim". Nevertheless, one could define a limiting average stochastic game by $\liminf_{N\to\infty}\frac{1}{N}\sum_{n=1}^{N}E_{s\pi_1\pi_2}(R_n(k))$. In Remark 2 of section 6 we indicate that our results also hold for this latter criterion.

Furthermore, the game is called a zero-sum stochastic game if $r^2(s,i,j) = -r^1(s,i,j)$ for all s,i,j.

A zero-sum game has a value $v \in \mathbf{R}^z$, if for all $\epsilon > 0$ there exist strategies $\pi_{1\epsilon}$ for player I and $\pi_{2\epsilon}$ for player II such that for all π_1 and π_2 and for all $s \in S$

$$e(s,\pi_{1\epsilon},\pi_2) \geq v_s - \epsilon \text{ and } e(s,\pi_1,\pi_{2\epsilon}) \leq v_s + \epsilon$$

where $e = v_\beta^1$ ($\beta \in [0,1)$) or $e = g^1$ corresponding to whether we observe the game as a β-discounted game or as a limiting average game.

The β-discounted value will be denoted by v_β; the limiting average value will be denoted by g. In case the value exists, strategies $\pi_{1\epsilon}$ and $\pi_{2\epsilon}$ are called ϵ-optimal strategies for player I and player II respectively. If we can take $\epsilon = 0$, then strategy π_{10} is called an optimal strategy for player I. A similar definition goes for player II.

In nonzero-sum games one is interested in ϵ-equilibria. An ϵ-equilibrium, for initial state $s \in S$, is a pair of strategies $(\pi_{1\epsilon},\pi_{2\epsilon})$ such that for all π_1 and π_2

$$e^1(s,\pi_1,\pi_{2\epsilon}) \leq e^1(s,\pi_{1\epsilon},\pi_{2\epsilon}) + \epsilon \text{ and}$$
$$e^2(s,\pi_{1\epsilon},\pi_2) \leq e^2(s,\pi_{1\epsilon},\pi_{2\epsilon}) + \epsilon,$$

where again $e = v_\beta$ ($\beta \in [0,1)$) or $e = g$; thus we speak of a β-discounted ϵ-equilibrium for initial state s, and of a limiting average ϵ-equilibrium for initial state s, respectively. If $(\pi_{1\epsilon},\pi_{2\epsilon})$ is an ϵ-equilibrium for all initial states, then $(\pi_{1\epsilon},\pi_{2\epsilon})$ is simply called an ϵ-equilibrium. A 0-equilibrium will be called an equilibrium. The interpretation of an ϵ-equilibrium is that, once players have "agreed" to play some specific ϵ-equilibrium, neither player can gain more than ϵ by changing his strategy, if his opponent does not change strategy. Thus, an ϵ-equilibrium is in a way self-enforcing: once it is agreed upon, it is not worthwhile to deviate from it.

2. Historical Review: Shapley [11], initiator of the theory of stochastic games, showed that zero-sum β-discounted stochastic games have a value and that both players possess optimal stationary strategies. For nonzero-sum β-discounted stochastic games Fink [4] showed the existence of equilibria consisting of stationary strategies.

Limiting average stochastic games appear to be more troublesome. Gillette [5] introduced them. Besides considering irreducible stochastic games, he gave an example for which it was not clear whether it had a value or not with respect to the limiting average criterion. The game was baptized "the big match" and it was first in 1968 that Blackwell and Ferguson [3] showed that behavior strategies, relying on the history of the play, are indispensable for even playing ϵ-optimal in the limiting average case. Later, Mertens and Neyman [7] were able to prove that all limiting average zero-sum stochastic games have a value. Their proof is based on the work of Bewley and Kohlberg [1] who showed the expansion, in fractional powers of $(1-\beta)$, of the β-discounted value for β close enough to 1.

Until now, the existence of ϵ-equilibria for limiting average nonzero-sum stochastic games has only been proved for several subclasses, i.e. stochastic games with special properties for the reward and/or transition functions. It was, for instance, Rogers [10] who showed the existence of stationary equilibria in irreducible stochastic games. Other classes, for which the existence of (ϵ-)equilibria has been shown, are for example: stochastic games in which one player controls the transitions (Parthasarathy and Raghavan [8]), stochastic games with state independent transitions and separable rewards (Parthasarathy et al. [9]), and repeated games with absorbing states (Vrieze and Thuijsman [13]).

In this paper we show that for all nonzero-sum stochastic games the set of states for which an ϵ-equilibrium exists, is non-empty. In fact we show that on some subset of states an almost stationary ϵ-equilibrium exists. Here "almost stationary" refers to the fact that each player uses a stationary strategy unless he has reasons to believe that the other player is deviating. In the latter case he will start playing some, possibly behavioral, punishment strategy.

3. Preliminaries: In this section we state some well-known facts about stochastic games, introduce some notations and derive some new basic results, needed in the sequel. For a pair of stationary strategies (τ,σ) we define:

- $\text{Car}(\tau) := \Pi_{s=1}^{z}\text{Car}(\tau_s)$, where $\text{Car}(\tau_s) := \{i \in A_s : \tau_s(i) > 0\}$ is the carrier of τ_s. $\text{Car}(\sigma)$ and $\text{Car}(\sigma_s)$ have a similar meaning.

- $r^k(\tau,\sigma) := (r^k(1,\tau_1,\sigma_1), r^k(2,\tau_2,\sigma_2),\ldots,r^k(z,\tau_z,\sigma_z))$ with $r^k(s,\tau_s,\sigma_s) := \sum_{i\in A_s}\sum_{j\in B_s}\tau_s(i)\sigma_s(j)r^k(s,i,j)$, being the expected immediate reward in state s to player $k\in\{1,2\}$, when player I uses τ_s and player II uses σ_s in state s.

- $P_s(\tau_s, \sigma_s) := (p(1|s, \tau_s, \sigma_s), p(2|s, \tau_s, \sigma_s), \ldots, p(z|s, \tau_s, \sigma_s))$ with $p(t|s, \tau_s, \sigma_s) := \sum_{i \in A_s} \sum_{j \in B_s} \tau_s(i)\sigma_s(j)p(t|s, i, j)$, being the probability to move from state s directly to state t if player I uses τ_s and player II uses σ_s.
- $P(\tau, \sigma)$:= the transition matrix for (τ, σ) with s-th row $P_s(\tau_s, \sigma_s)$ for $s = 1, 2, \ldots, z$.
  $P(\tau, \sigma)$ is a stochastic matrix, i.e. non-negative with rowsums 1. With $P(\tau, \sigma)$ we associate a Markov chain on the set of states, where the transition probability from state s to state t is given by entry (s, t) of $P(\tau, \sigma)$, being $p(t|s, \tau_s, \sigma_s)$.
- $Q(\tau, \sigma) := lim_{N \to \infty} \frac{1}{N} \sum_{n=1}^{N} (P(\tau, \sigma))^n$, being the Cesaro limit of $P(\tau, \sigma)$. $Q(\tau, \sigma)$ is a stochastic matrix which satisfies $Q(\tau, \sigma)P(\tau, \sigma) = Q(\tau, \sigma) = P(\tau, \sigma)Q(\tau, \sigma)$.

The β-discounted income, $\beta \in [0, 1)$, of player $k \in \{1, 2\}$, under (τ, σ) is the unique solution $v_\beta^k(\tau, \sigma) \in \mathbf{R}^z$ of the Shapley equation

$$v_\beta^k(\tau, \sigma) = r^k(\tau, \sigma) + \beta P(\tau, \sigma) v_\beta^k(\tau, \sigma). \tag{1}$$

Since $(I - \beta P(\tau, \sigma))^{-1}$ always exists for $\beta \in [0, 1)$, the β-discounted income of player k can also be given as

$$v_\beta^k(\tau, \sigma) = (I - \beta P(\tau, \sigma))^{-1} r^k(\tau, \sigma). \tag{2}$$

The limiting average income of player $k \in \{1, 2\}$ under (τ, σ) is given by

$$g^k(\tau, \sigma) := \lim_{N \to \infty} \frac{1}{N} \sum_{n=1}^{N} (P(\tau, \sigma))^n r^k(\tau, \sigma) = Q(\tau, \sigma) r^k(\tau, \sigma).$$

By Bl^^kwell [2] it holds that $g^k(\tau, \sigma) = lim_{\beta \to 1}(1 - \beta) v_\beta^k(\tau, \sigma)$ and thus we have by (2)

$$g^k(\tau, \sigma) = Q(\tau, \sigma) r^k(\tau, \sigma) = \lim_{\beta \to 1} (1 - \beta)(I - \beta P(\tau, \sigma))^{-1} r^k(\tau, \sigma). \tag{3}$$

By Mertens and Neyman [7] it is also well-known that the β-discounted values v_β and the limiting average value g of a zero-sum stochastic game are related by

$$g = \lim_{\beta \to 1} (1 - \beta) v_\beta. \tag{4}$$

With any nonzero-sum stochastic game we associate two zero-sum stochastic games, called the 1 zero-sum and 2 zero-sum stochastic game respectively. Here the k zero-sum stochastic game, $k \in \{1, 2\}$, is obtained by deleting the payoffs of player $3-k$ and considering the zero-sum game with payoffs $r^k(s,i,j)$ to the maximizing player k and $-r^k(s,i,j)$ to the minimizing player $3-k$. The values of the k zero-sum game, $k \in \{1,2\}$, will be denoted by v_β^k and g^k for the β-discounted and the limiting average criterion respectively.

By Fink [4] there exists a stationary β-discounted equilibrium for all $\beta \in [0,1)$. Knowing this, we can consider a sequence of stationary β-discounted equilibria for β tending to 1. Such a sequence will be the starting point of our investigations.

So, let for each $n \in \{1,2,3,\ldots\}$, $\beta_n \in [0,1)$ and $lim_{n\to\infty} \beta_n = 1$, such that $(\tau^{\beta_n}, \sigma^{\beta_n})$ is a stationary β_n-discounted equilibrium for all n. Without loss of generality, using compactness arguments, we assume that the following limits exist and we define:

d1: $(\tau^1, \sigma^1) := lim_{n\to\infty}(\tau^{\beta_n}, \sigma^{\beta_n})$

d2: For $k=1,2$: $G^k := lim_{n\to\infty}(1-\beta_n)v_{\beta_n}^k(\tau^{\beta_n}, \sigma^{\beta_n})$

d3: $X^{\beta_n} := (1-\beta_n)(I-\beta_n P(\tau^{\beta_n}, \sigma^{\beta_n}))^{-1}$ and $X^1 := lim_{n\to\infty} X^{\beta_n}$.

Notice that (d1) implies that $P(\tau^1, \sigma^1) = lim_{n\to\infty} P(\tau^{\beta_n}, \sigma^{\beta_n})$.

From now on we suppress the index n and write $lim_{\beta\to 1}(\tau^\beta, \sigma^\beta) = (\tau^1, \sigma^1)$ instead of (d1) and likewise for the other limits. We also assume, without loss of generality, that $\mathrm{Car}(\tau^\beta)$ and $\mathrm{Car}(\sigma^\beta)$ are both independent of $\beta \in [0\,1)$ for this sequence. This implies that the transition structure of the related Markov chain does not depend on $\beta \in [0\,1)$.

The following lemma follows straightforward from the definition of equilibrium.

Lemma 1: Let $(\tau^\beta, \sigma^\beta)$ be a stationary β-discounted equilibrium. Then:

(i) For each τ with $\mathrm{Car}(\tau) \subset \mathrm{Car}(\tau^\beta)$ and for each σ with $\mathrm{Car}(\sigma) \subset \mathrm{Car}(\sigma^\beta)$:

$$v_\beta^1(\tau^\beta, \sigma^\beta) = v_\beta^1(\tau, \sigma^\beta) \text{ and } v_\beta^2(\tau^\beta, \sigma^\beta) = v_\beta^2(\tau^\beta, \sigma)$$

(ii) For all τ and σ (componentwise):

$$\begin{aligned} v_\beta^1(\tau^\beta, \sigma^\beta) &= r^1(\tau^\beta, \sigma^\beta) + \beta P(\tau^\beta, \sigma^\beta) v_\beta^1(\tau^\beta, \sigma^\beta) \\ &\geq r^1(\tau, \sigma^\beta) + \beta P(\tau, \sigma^\beta) v_\beta^1(\tau^\beta, \sigma^\beta); \\ v_\beta^2(\tau^\beta, \sigma^\beta) &= r^2(\tau^\beta, \sigma^\beta) + \beta P(\tau^\beta, \sigma^\beta) v_\beta^2(\tau^\beta, \sigma^\beta) \\ &\geq r^2(\tau^\beta, \sigma) + \beta P(\tau^\beta, \sigma) v_\beta^2(\tau^\beta, \sigma^\beta). \end{aligned}$$

(iii) $v_\beta^k(\tau^\beta, \sigma^\beta) \geq v_\beta^k, \quad k \in \{1,2\}$.

By (4), (d2) and Lemma 1(iii) one can derive

$$G^k \geq g^k, \text{ for } k = 1,2. \tag{5}$$

The next lemma follows by multiplying the equations in Lemma 1(ii) by $(1-\beta)$ and applying Lemma 1(i) and (d2).

Lemma 2: For each $\tilde{\tau}$ with $\mathrm{Car}(\tilde{\tau}) \subset \mathrm{Car}(\tau^\beta)$ and for all τ;

$$G^1 = P(\tau^1, \sigma^1)G^1 = P(\tilde{\tau}, \sigma^1)G^1 \geq P(\tau, \sigma^1)G^1.$$

A similar statement holds for strategies $\tilde{\sigma}$ and σ for player II.

Let $S^1, S^2, \ldots, S^H$ be the ergodic classes and let T be the set of transient states, both with respect to (τ^1, σ^1).

The following lemma says that G^k_s is constant per ergodic class S^h, $h \in \{1, 2, \ldots, H\}$.

Lemma 3: $G^k_s = G^k_t =: G^{kh}$ for $s, t \in S^h$ and $h \in \{1, 2, \ldots, H\}$.

Proof: From Lemma 2 we have $G^k(h) = P^h(\tau^1, \sigma^1)G^k(h)$, where $G^k(h)$ is defined as the restriction of G^k to the states S^h and where $P^h(\tau^1, \sigma^1)$ is the restriction of $P(\tau^1, \sigma^1)$ to the states of S^h. Since $P^h(\tau^1, \sigma^1)$ is irreducible, this equality can only hold in case each coordinate of $G^k(h)$ is the same. ◇

The next result, which is new, is of crucial importance for our proof:

Lemma 4: $X^1 P(\tau^1, \sigma^1) = X^1$.

Proof: Since $X^\beta(I - \beta P(\tau^\beta, \sigma^\beta)) = (1 - \beta)I$, we have $\beta X^\beta P(\tau^\beta, \sigma^\beta) = X^\beta - (1 - \beta)I$, which gives the assertion in the lemma, when taking limits. ◇

Notice that for all β, X^β is a stochastic matrix, hence also X^1.

By rearranging the states, $P(\tau^1, \sigma^1)$ can be reduced to

$$P(\tau^1, \sigma^1) = \begin{bmatrix} P^1(\tau^1,\sigma^1) & 0 & \cdots & 0 & 0 \\ 0 & P^2(\tau^1,\sigma^1) & & 0 & 0 \\ & & \ddots & & \\ 0 & 0 & & P^H(\tau^1,\sigma^1) & 0 \\ P^{T1}(\tau^1,\sigma^1) & P^{T2}(\tau^1,\sigma^1) & \cdots & P^{TH}(\tau^1,\sigma^1) & P^T(\tau^1,\sigma^1) \end{bmatrix}$$

Here $P^h(\tau^1, \sigma^1)$ corresponds to the h-th irreducible class, S^h, with respect to (τ^1, σ^1), for $h = 1, 2, \ldots, H$. The bottom rows correspond with transitions from transient states. Obviously, there exists at least one irreducible class, but the set of transient states, T, may be empty. It is well-known that $(I - P^T(\tau^1, \sigma^1))^{-1}$ exists.

Now $Q(\tau^1,\sigma^1)$ has the following shape (cf. Kemeny and Snell [6]).

$$Q(\tau^1,\sigma^1) = \begin{bmatrix} Q^1(\tau^1,\sigma^1) & 0 & \cdots & 0 & 0 \\ 0 & Q^2(\tau^1,\sigma^1) & & 0 & 0 \\ & & \ddots & & \\ 0 & 0 & & Q^H(\tau^1,\sigma^1) & 0 \\ Q^{T1}(\tau^1,\sigma^1) & Q^{T2}(\tau^1,\sigma^1) & \cdots & Q^{TH}(\tau^1,\sigma^1) & 0 \end{bmatrix}$$

Here for $h = 1,2,\ldots,H$, $Q^h(\tau^1,\sigma^1)$ is the Cesaro limit of $P^h(\tau^1,\sigma^1)$ and

$$Q^{Th}(\tau^1,\sigma^1) = (I - P^T(\tau^1,\sigma^1))^{-1}P^{Th}(\tau^1,\sigma^1)Q^h(\tau^1,\sigma^1).$$

For each $h = 1,2,\ldots,H$, the rows of $Q^h(\tau^1,\sigma^1)$ are identical and all equal the unique invariant distribution of the irreducible Markov chain corresponding to $P^h(\tau^1,\sigma^1)$. Furthermore, each entry of $Q^h(\tau^1,\sigma^1)$ is strictly positive.

Let $q^h := (0,0,\ldots,0,q_1^h,q_2^h,\ldots,q_{z_h}^h,0,\ldots,0) \in \mathbf{R}^z$, where $(q_1^h,q_2^h,\ldots,q_{z_h}^h)$ is the unique invariant distribution of $P^h(\tau^1,\sigma^1)$, and $z_h := |S^h|$.

Lemma 5: For each $s \in S$, let X_s^1 denote the s-th row of X^1.

(i) For each $s \in S$, there exist $\mu_s^1,\mu_s^2,\ldots,\mu_s^H \in [0,1]$ with $\sum_{h=1}^H \mu_s^h = 1$, such that $X_s^1 = \sum_{h=1}^H \mu_s^h q^h$.

(ii) If, for $h \in \{1,2,\ldots,H\}$, the ergodic class S^h cannot be reached from state s under $(\tau^\beta,\sigma^\beta)$, then $\mu_s^h = 0$.

Proof:

(i) By Lemma 4, each X_s^1 corresponds to a stationary distribution of the Markov chain induced by $P(\tau^1,\sigma^1)$. The set of stationary distributions of that Markov chain is exactly the convex hull of $\{q^1,q^2,\ldots,q^H\}$.

(ii) If S^h cannot be reached from s for each β, then, for all $n \in \mathbf{N}$ and $t \in S^h$, the (s,t)-th entry of $(P(\tau^\beta,\sigma^\beta))^n$ is 0. Hence this also holds for the (s,t)-th entry of X^β and thus also for X^1. ◇

Notice that $g^k(s,\tau^1,\sigma^1) = g^k(t,\tau^1,\sigma^1)$ for $s,t \in S^h$, which number will be denoted by $g^{kh}(\tau^1,\sigma^1)$.

Lemma 6: For $k = 1, 2$ and $s \in S$ let μ_s^h be as in Lemma 5. Then:

$$G_s^k = \sum_{h=1}^{H} \mu_s^h g^{kh}(\tau^1, \sigma^1).$$

Proof: Let $\langle , \rangle$ denote the innerproduct. We have:

$$\begin{aligned} G_s^k &= \lim_{\beta \to 1} (1-\beta) v_\beta^k(s, \tau^\beta, \sigma^\beta) \\ &= \lim_{\beta \to 1} \langle X_s^\beta, r^k(\tau^\beta, \sigma^\beta) \rangle \\ &= \langle X_s^1, r^k(\tau^1, \sigma^1) \rangle \\ &= \sum_{h=1}^{H} \mu_s^h \langle q^h, r^k(\tau^1, \sigma^1) \rangle = \sum_{h=1}^{H} \mu_s^h g^{kh}(\tau^1, \sigma^1). \end{aligned}$$

◇

From this lemma we can directly conclude the following:

Corollary 7: There exists an $h_1 \in \{1, 2, \ldots, H\}$ for which $g^{1h_1}(\tau^1, \sigma^1) \geq max_{s \in S}\, G_s^1$ and there exists an $h_2 \in \{1, 2, \ldots, H\}$ for which $g^{2h_2}(\tau^1, \sigma^1) \geq max_{s \in S}\, G_s^2$.

Remark: Lemma 3 to Corollary 7 remain true, even if the strategy pairs $(\tau^\beta, \sigma^\beta)$ are no β-discounted equilibria.

4. The zero-sum case: The following theorem has already been proved by Tijs and Vrieze [12]. However, their proof is rather complex, whereas the proof given below is easy; it follows almost directly from the above results.

Theorem 8: For every zero-sum limiting average stochastic game, there is for each player a set of initial states for which an optimal stationary strategy exists.

Proof: We show that player I has an optimal stationary strategy on some subset of states.

Let τ^β be a β-discounted optimal stationary strategy for player I in the β-discounted stochastic game. Without loss of generality, we assume that $lim_{\beta\to1}\tau^\beta$ exists, and we define $\tau^1 := lim_{\beta\to1}\tau^\beta$.

It is well-known that against τ^1 player II has a stationary best answer, i.e. there is some stationary strategy σ such that for all strategies π_2 it holds that $g(\tau^1,\sigma) \leq g(\tau^1,\pi_2)$.

By the remark at the end of section 3, focussing on player I while taking $\sigma^\beta = \sigma$ for all β, Corollary 7 gives us that there exists an $h \in \{1,2,\dots,H\}$ for which $g^h(\tau^1,\sigma) \geq max_{s\in S}\, G_s \geq max_{s\in S}\, g_s$. Hence, for all $s \in S^h$ it holds that $g(s,\tau^1,\sigma) \geq g_s$.

Since σ is a best reply against τ^1, it follows that for all $s \in S^h$ and for all strategies π_2 we have: $g(s,\tau^1,\pi_2) \geq g_s$, which means that τ^1 is limiting average optimal for player I, for all initial states in S^h. ◇

5. The nonzero-sum case:

Theorem 9: For every nonzero-sum limiting average stochastic game, there is some subset of states on which an almost stationary ϵ-equilibrium exists.

This theorem follows directly from Lemma 10 and Lemma 11 below.

Lemma 10: Let $h \in \{1,2,\dots,H\}$ be such that $g^{kh}(\tau^1,\sigma^1) \geq G^{kh}$ for $k = 1,2$. Then an ϵ-equilibrium for initial states in S^h can be made by supplementing τ^1 and σ^1 with suitable threats (for punishment) to prevent deviations.

Proof: When a player deviates, say player I, with some action i outside $\text{Car}(\tau_s^1)$ in some state $s \in S^h$, then by Lemma 2 and Lemma 3:

$$G_s^1 = G^{1h} \geq \sum_{t=1}^{z} p(t|s,i,\sigma_s^1)G_t^1. \tag{6}$$

Since player II can observe such a deviation, he can counter at once by playing $\epsilon/2$-optimal in the 1 zero-sum stochastic game (cf. section 3) from that moment on, giving player I a limiting average reward of at most (cf. (5) and (6))

$$\sum_{t\in S} p(t|s,i,\sigma_s^1)(g_t^1 + \epsilon/2) \leq G^{1h} + \epsilon/2. \tag{7}$$

Similarly, deviations of player II outside $\text{Car}(\sigma_s^1)$ can be countered. Notice that (6), with equality sign, and (7) also hold for actions i inside $\text{Car}(\tau_s^1)$.

Deviations inside $\mathrm{Car}(\tau_s^1)$ cannot be observed immediately. However, if both players really use τ^1 and σ^1 on S^h then the state action frequency of action i in state s will approach $q_s^h \tau_s^1(i)$ with probability 1, for each $i \in \mathrm{Car}(\tau_s^1)$ and each $s \in S^h$. (Remember that $(q_1^h, \ldots, q_{z_h}^h)$ is the invariant distribution corresponding to (τ^1, σ^1) on S^h).

Let $Y_s^n(h)$ denote the stochastic variable of the state action frequency up to stage n, $s \in S^h$, $i \in \mathrm{Car}(\tau_s^1)$, and let $y_s^n(i)$ be the observed realization of $Y_s^n(i)$. Then for every $\delta > 0$ and $\alpha > 0$ there exists $N_{\delta\alpha} \in \mathbf{N}$ such that

$$Pr\left\{ \sup_{n \geq N_{\delta\alpha}, i \in A_s, s \in S^h} |Y_s^n(i) - q_s^h \tau_s^1(i)| > \alpha, \text{ given player I uses } \tau^1 \right.$$
$$\left. \text{and player II uses } \sigma^1 \text{ on } S^h \right\} < \delta.$$

Notice that if $|y_s^n(i) - q_s^h \tau_s^1(i)| \leq \alpha$ for each $n \geq N_{\delta\alpha}$, then the limiting average reward is

$$\text{at least } g^{kh}(\tau^1, \sigma^1) - \alpha M \text{ and at most } g^{kh}(\tau^1, \sigma^1) + \alpha M \tag{8}$$

(M is the maximal absolute direct payoff.)

Now let $\alpha < \epsilon/2M$ and choose δ so small that for $k = 1, 2$:

$$(1 - \delta) g^{kh}(\tau^1, \sigma^1) - \delta M \geq g^{kh}(\tau^1, \sigma^1) - \epsilon/2. \tag{9}$$

Then player II can suppress incentives of player I to deviate from τ_s^1, $s \in S^h$, in the following way: Keep playing σ^1 unless $|y_s^n(i) - q_s^h \tau_s^1(i)| > \alpha$ for some $s \in S^h$, $i \in \mathrm{Car}(\tau_s^1)$ and $n \geq N_{\delta\alpha}$. If that happens, start playing an $\epsilon/2$-optimal strategy in the 1 zero-sum stochastic game.

If player I does not deviate, then with probability at most δ punishment occurs. However, because of (9), in this case the limiting average reward to the players is

$$\text{at least } g^{kh}(\tau^1, \sigma^1) - \epsilon/2, \; k = 1, 2. \tag{10}$$

If player I does deviate inside $\mathrm{Car}(\tau_s^1)$, which may or may not cause punishment, then, by (8), (7) and the punishment strategy, the limiting average reward to player I is

$$\text{at most } g^{1h}(\tau^1, \sigma^1) + \epsilon/2. \tag{11}$$

By combining (7), (10), and (11) it follows that player I can gain at most ϵ by unilateral deviating from τ^1.

For player I similar threats for punishment in case of deviations of player II can be defined. In the combined case δ must be chosen a bit more sensitive to ensure that undeserved punishment of one of the players lowers the rewards by at most $\epsilon/2$ (compare (9)). $\diamond$

Lemma 11: There exist $h_1, h_2 \in \{1,2,\dots,H\}$ such that

(a) $g^{1h_1}(\tau^1,\sigma^1) \geq max_{s\in S}\, G_s^1 \geq G^{1h_1}$ and $g^{2h_1}(\tau^1,\sigma^1) \geq G^{2h_1}$.

(b) $g^{1h_2}(\tau^1,\sigma^1) \geq G^{1h_2}$ and $g^{2h_2}(\tau^1,\sigma^1) \geq max_{s\in S}\, G_s^2 \geq G^{2h_2}$.

Proof: Let T be the set of transient states w.r.t. (τ^1,σ^1). For a subset D of $\{1,2,\dots,H\}$ let $S(D) := \bigcup_{h\in D} S^h$.

Define $D_1 := \{h \in \{1,2,\dots,H\} : g^{1h}(\tau^1,\sigma^1) \geq max_{s\in S}\, G_s^1\}$. Corollary 7 implies $D_1 \neq \emptyset$. Define $T_1 := \{s \in T$: starting in s, under (τ^1,σ^1) the play will ever reach $S(D_1)$ with probability 1$\}$.

The proof is divided into two parts. In part 1 we show that there exists a non-empty subset D_2 of D_1 and a subset T_2 of T_1 with the property that under (τ^β,σ^1) the set $S(D_2)\cup T_2$ is irreducible. In part 2 we show that for some $h\in D_2$ it holds that $g^{2h}(\tau^1,\sigma^1) \geq G^{2h}$.

Part 1 Let τ be a stationary strategy for player I such that $\text{Car}(\tau_s) = \text{Car}(\tau_s^\beta)$ for all $s \in S(D_1)$ and $\tau_s = \tau_s^1$ for $s \in S\backslash S(D_1)$. Then each irreducible set of states with respect to (τ,σ^1) is either a subset of $S(D_1)\cup T_1$, or one of the classes S^h, $h \in \{1,2,\dots,H\}\backslash D_1$. We show that there exists an irreducible set of states within $S(D_1)\cup T_1$, w.r.t. (τ,σ^1).

Suppose that all states in $S(D_1)$ are transient w.r.t. (τ,σ^1). This gives that $D_1 \neq \{1,2,\dots,H\}$, otherwise all states of S would be transient w.r.t. (τ,σ^1). Hence $\{S^h : h\in D_1^c\}$ is non-empty and contains, by definition of τ, all irreducible classes of (τ,σ^1). By Lemma 1(i) and analogues of Lemma 5 and Lemma 6 we derive that for all $t\in S$:

$$G_t^1 = \lim_{\beta\to 1}(1-\beta)v_\beta^1(t,\tau^\beta,\sigma^\beta) = \lim_{\beta\to 1}(1-\beta)v_\beta^1(t,\tau,\sigma^\beta) =$$
$$= \sum_{h\in D_1^c} \mu_t^h g^{1h}(\tau,\sigma^1) = \sum_{h\in D_1^c} \mu_t^h g^{1h}(\tau^1,\sigma^1) < \max_s G_s^1,$$

for some $\mu_t^h \geq 0$ for all h, t and $\sum_{h\in D_1^c}\mu_t^h = 1$, which is, of course, a contradiction. Hence there exists a non-empty subset D_2 of D_1 and a subset T_2 of T_1 as desired.

Part 2 Again applying Lemma 1(i) and analogues of Lemmas 5 and 6, we find that for all $t\in S(D_2)$:

$$G_t^2 = \lim_{\beta\to 1}(1-\beta)v_\beta^2(t,\tau^\beta,\sigma^\beta) = \lim_{\beta\to 1}(1-\beta)v_\beta^2(t,\tau^\beta,\sigma^1) =$$
$$= \sum_{h\in D_2} \mu_t^h g^{2h}(\tau^1,\sigma^1) \leq \max_{h\in D_2} g^{2h}(\tau^1,\sigma^1)$$

for some $\mu_t^h \geq 0$ for all h, t and $\sum_{h\in D_2}\mu_t^h = 1$. Hence for at least one $h\in D_2$ it holds that $g^{2h}(\tau^1,\sigma^1) \geq G^{2h}$. ◇

Notice that if we let $E := \{h \in \{1,2,\ldots,H\} : g^{kh}(\tau^1,\sigma^1) \geq G^{kh}$, for $k = 1,2\}$. Then E is non-empty by Lemma 11. By Lemma 10 (τ^1,σ^1) supplemented with threats forms an ϵ-equilibrium for initial states in S^h, $h \in E$, and if the players stick to this ϵ-equilibrium then with probability at least $(1-\delta)^2$ the play stays within E forever.

6. Remarks:

Remark 1: Our proof can easily be extended to N-player stochastic games. The Lemmas 1-7 straightforwardly hold for the N-player case as well. For Lemma 10 the concept of k zero-sum stochastic game (cf. section 3) has to be adapted. For the N-player case the associated k zero-sum stochastic game is defined as the game where only the payoffs of player k play a role. Player k tries to maximize his reward (discounted or limiting average, whatever the criterion) while the remaining $N-1$ players jointly try to minimize this reward. Notice the difference with an ordinary zero-sum stochastic game. The minimizing "player" chooses at each stage an $(N-1)$-tuple of joint actions instead of one action. It can be shown that also for this class of stochastic games the Mertens and Neyman approach goes through and that the limiting average value equals the limit of $(1-\beta)$ times the β-discounted value for β tending to 1. Then in Lemma 10 the threats can be adapted in the following way: if player k deviates, no matter how, then the other $N-1$ players punish jointly by playing $\epsilon/2$-optimal in the k zero-sum stochastic game, etc.

The proof of Lemma 11 can be extended by successively treating all players: there exists a non-empty subset D_3 of D_2 for which the desired property for player 3 holds, etc.

Remark 2: As mentioned in section 1 a limiting average stochastic game could be defined by the alternative criterion

$$a^k(s,\pi_1,\pi_2) = \liminf_{N\to\infty} \frac{1}{N}\sum_{n=1}^{N} E_{s\pi_1\pi_2}(R_n(k)).$$

Theorems 8 and 9 hold for this alternative criterion as well. This follows by noting that:

i) Mertens and Neyman [7] have shown that for the zero-sum case the value is the same for the limiting average criterion and the above criterion, and furthermore, there do exist strategies that are ϵ-optimal with respect to both criteria at the same time.

ii) Let $(\pi_1(\epsilon),\pi_2(\epsilon))$ be an ϵ-equilibrium (constructed as in our proof) for the original criterion using punishment strategies as mentioned in (i). By Fatou's lemma it holds that $a^k(s,\pi_1(\epsilon),\pi_2(\epsilon)) \geq g^k(s,\pi_1(\epsilon),\pi_2(\epsilon))$, and we already know that $g^k(s,\pi_1(\epsilon),\pi_2(\epsilon)) \geq G^k_s - \epsilon/2$.

iii) Suppose player II uses some strategy π_2 against $\pi_1(\epsilon)$. Let $s \in S^h$ be an easy initial state. Under the condition of punishment

$$a^2(s, \pi_1(\epsilon), \pi_2) \leq g_s^2 + \epsilon/2 \leq g^2(s, \pi_1(\epsilon), \pi_2(\epsilon)) + \epsilon/2 \leq$$
$$a^2(s, \pi_1(\epsilon), \pi_2(\epsilon)) + \epsilon/2.$$

Under the condition of no-punishment the state-action frequencies $y_t^n(j)$ (as defined in the proof of Lemma 10) according to π_2 satisfy $|y_t^n(j) - q_t^h \sigma_t^1(j)| \leq \alpha$ for all $n \geq N_{\delta\alpha}$ and $t \in S^h$. This implies that for n sufficiently large

$$\left| \frac{1}{N} \sum_{n=1}^{N} E_{s\pi_1(\epsilon)\pi_2}(R_n(2)) - \frac{1}{N} \sum_{n=1}^{N} E_{s\pi_1(\epsilon)\pi_2(\epsilon)}(R_n(2)) \right| \leq \epsilon/2 \text{ and so}$$
$$a^2(s, \pi_1(\epsilon), \pi_2) \leq a^2(s, \pi_1(\epsilon), \pi_2(\epsilon)) + \epsilon/2.$$

Remark 3: Note that we explicitly use the result of Mertens and Neyman [7] for zero-sum limiting average stochastic games.

Remark 4: Our results imply that stationary equilibria exist for irreducible stochastic games. Hence we have a new proof for this result of Rogers [10]. Irreducible stochastic games are stochastic games with the property that $P(\tau, \sigma)$ is irreducible for each pair of stationary strategies (τ, σ).

The existence of stationary equilibria follows because:

i) S is the only irreducible set for all pairs of stationary strategies.

ii) Let $(\tau^\beta, \sigma^\beta)$ be a sequence of β-discounted stationary equilibria as in section 3, and let σ be some arbitrary stationary strategy for player II.

iii) Having (i) and (ii), Lemma 6 and the remark at the end of section 3 imply that for all states $s \in S$:

$$g^2(s, \tau^1, \sigma) = \lim_{\beta \to 1} (1-\beta) v_\beta^2(s, \tau^\beta, \sigma) \leq \lim_{\beta \to 1} (1-\beta) v_\beta^2(s, \tau^\beta, \sigma^\beta) = g^2(s, \tau^1, \sigma^1).$$

iv) When playing against a stationary strategy, player II has a stationary best reply. Hence it follows from (iii) that $g^2(\tau^1, \pi_2) \leq g^2(\tau^1, \sigma^1)$ for all strategies π_2.

References

[1] *Bewley, T. and E. Kohlberg* (1976). The Asymptotic Theory of Stochastic Games, *Math. Oper. Res.*, **1**, 197-208.

[2] *Blackwell, D.* (1962). Discrete Dynamic Programming, *Ann. of Math. Stat.*, **33**, 719-726.

[3] *Blackwell, D. and T.S. Ferguson* (1968). The Big Match, *Ann. of Math. Stat.*, **39**, 159-163.

[4] *Fink, A.M.* (1964). Equilibrium in a Stochastic n-Person Game, *J. of Sci. of Hiroshima Univ.*, Series A-I **28**, 89-93.

[5] *Gillette, D.* (1957). Stochastic Games with Zero Stop Probabilities, In: M. Dresher, A.W. Tucker and P. Wolfe (eds.), *Contributions to the theory of games*, vol. II, *Annals of Math. Studies*, **39**, Princeton, 197-208.

[6] *Kemeny, J. and J. Snell* (1961). Finite Markov Chains, Van Nostrand, Princeton.

[7] *Mertens, J.F. and A. Neyman* (1981). Stochastic Games, *Int. J. Game Theory*, **10**, 53-66.

[8] *Parthasarathy, T. and T.E.S. Raghavan* (1981). An Orderfield Property for Stochastic Games when One Player Controls Transition Probabilities, *J. Opt. Theory Appl.*, **33**, 375-392.

[9] *Parthasarathy, T., S.H. Tijs, and O.J. Vrieze* (1984). Stochastic Games with State Independent Transitions and Separable Rewards, In: G. Hammer and D. Pallaschke (eds.), *Selected Topics in Operations Research and Mathematical Economics*, Springer-Verlag, Berlin, 262-271.

[10] *Rogers, P.D.* (1969). Nonzero-Sum Stochastic Games, Ph.D. Dissertation, Report ORC 69-8, Op. Res. Centre, Univ. of California, Berkeley.

[11] *Shapley, L.S.* (1953). Stochastic Games, *Proc. of the Nat. Acad. of Sc., U.S.A.*, **39**, 1095-1100.

[12] *Tijs, S.H. and O.J. Vrieze* (1986). On the Existence of Easy Initial States for Undiscounted Stochastic Games, *Math. Oper. Res.*, **11**, 506-513.

[13] *Vrieze, O.J. and F. Thuijsman* (1989). On Equilibria in Repeated Games with Absorbing States, Int. J. Game Theory, **18**, 293- 310.

F. Thuijsman
Department of Mathematics
Faculty of General Sciences
University of Limburg
P.O. Box 616
6200 MD Maastricht
The Netherlands

O.J. Vrieze
Department of Mathematics
Faculty of General Sciences
University of Limburg
P.O. Box 616
6200 MD Maastricht
The Netherlands

"Big Match" with Lack of Information on One Side (III)

Sylvain Sorin and Shmuel Zamir

Abstract: We prove the existence of a minmax for a class of stochastic games with incomplete information on one side by using an auxiliary one shot game.

1. Introduction: We study here a new specific class of two person zero-sum stochastic games with lack of information on one side; we assume independence and one non absorbing state. The first hypothesis means that the state space can be represented as $K \times L$, the initial probability being on K (k is fixed from then on) and the transition on L being independent of k. Hence the game can be viewed as a "stochastic game with vector payoffs". (The second assumption is now clear).

A first approach appeared in Sorin [8], [9] studying "Big Match"- type games (i.e. where one line or one column is absorbing, see Blackwell and Ferguson [2]) but preliminary analysis of some other cases may still be needed before getting an idea of the general kind of strategies involved.

However interesting properties and conjectures are already available:

First, in the two papers quoted above, it was observed that $max\, min = lim\, v_n$, i.e. that the informed player could do as well in the infinitely repeated game than in long finite games and it was conjectured that this relation holds true in all stochastic games with lack of information on one side. In fact it is probably true in a much more general framework, namely repeated games with the following property (P): player 1 is more informed than player 2. This follows from two recent results of Mertens [4], see also Melolidakis [3]:

i) Given any general repeated game there exists a natural compact state space where the original game can be represented as a stochastic game such that finite and discounted values coincide.

ii) Moreover under (P) one can mimic strategies of player 1 (and best replies of player 2) in the stochastic game to get similar properties in the original game. Assuming the stochastic game to have a value (which seems to be a reasonable conjecture) one then obtains the existence of $max\,min$, $lim\, v_n$ and $lim\, v_\lambda$ in the original game and their equality.

These facts explain why we will be concerned with the other aspect of the problem, namely existence and characterization of the minmax. As in the previous articles the minmax will appear as the value of a (one shot) auxiliary game with lack of information on one side: somehow the infinitely repeated game can be "normalized" when looking at strategies of player 2 and best replies of player 1. Note that similar tools were introduced and used by Mertens and Zamir [6], see also Waternaux [10], [11], to study games with incomplete information and no signals.

Finally we will provide (when there are two states of nature) a simple geometric characterization of the minmax based on the orthants that player 2 can approach in the game with vector payoffs and absorbing states. The aim is obviously to get, in this framework, a dual (geometric/analytic) necessary and sufficient condition for approachability of convex sets, as follows from Blackwell [1] in the "deterministic" case.

2. The Game: Let $G^k, k \in K$ be a finite set of 2×2 payoff matrices of the form $G^k = \begin{pmatrix} a^{*k} & b^k \\ c^k & d^k \end{pmatrix}$, where the star $(*)$ denotes an absorbing payoff. We can assume $|G^k| \leq 1, \forall k \in K$. For each p, probability on K,$G(p)$ denotes the infinitely repeated game with the following common knowledge rules : k is chosen according to p, player 1 (the row player) is informed about k, players choose their moves simultaneously (full monitoring is assumed : at each stage, both players are told the previous moves) and the stage payoff is computed according to G^k.

Denote by I and J the set of moves of player 1 and player 2. As usual histories are finite sequences of couples in $I \times J$ and write H for the corresponding set. Strategies for player 1 (resp. 2) are transition probabilities from $K \times H$ to I (resp. from H to J) and are denoted by σ in Σ (resp. $\tau \in \mathcal{T}$). A couple (σ, τ) and k (or equivalently σ^k and τ) defines a probability on the product σ-algebra on the set of plays $H_\infty = (I \times J)^\infty$. To each play corresponds a stream of vector payoffs $(g_1, \ldots, g_n, \ldots)$,$\overline{g}_n$ denotes the average up to stage n and $\overline{\gamma}^p_n(\sigma, \tau)$ the corresponding expected payoff under p, σ and τ. We thus have: $\overline{\gamma}^p_n(\sigma, \tau) = \sum_k p^k \overline{\gamma}^k_n(\sigma^k, \tau)$ and $\overline{\gamma}^k_n(\sigma^k, \tau) = E_{\sigma^k, \tau}(\overline{g}^k_n)$.

Finally $\overline{v}(p)$ is the minmax of $G(p)$ if:

(i) $\forall \varepsilon > 0, \exists N, \exists \tau, \forall n \geq N, \forall \sigma, \qquad \overline{\gamma}^p_n(\sigma, \tau) \leq \overline{v}(p) + \varepsilon$

(ii) $\forall \varepsilon > 0, \forall \tau, \exists N, \exists \sigma, \forall n \geq N, \qquad \overline{\gamma}^p_n(\sigma, \tau) \geq \overline{v}(p) - \varepsilon.$

Remark: The value of the game $G(k)$ (corresponding to p concentrated on k) is easily seen to be:

$$\text{if } a^k \geq d^k : \quad \text{med } (a^k, b^k, d^k)$$
$$\text{if } a^k \leq d^k : \quad \text{med } (a^k, c^k, d^k)$$

and ε-optimal strategies can be chosen in the following set (say for player 1): always Top ($\widetilde{T}$), always Bottom ($\widetilde{B}$), always $(x, 1-x)$ i.i.d. with $x \in (0,1)$.

3. An Auxiliary Game: $\Gamma(p)$ is the (one shot) game in normal form defined by X^K (resp. Y) strategy set of player 1 (resp. 2) and payoff f where: $X = \overline{\mathcal{N}} \cup \{\partial\}$ ($\overline{\mathcal{N}}$ is the compactification $\mathcal{N} \cup \{\infty\}$ of the set of positive integers $\mathcal{N}$ and ∂ is some isolated point with $\partial > \infty$)

$$Y = \{0,1\}^{\mathcal{N}}$$

the payoff f is the average of the state payoff f^k,

$$f(x,y) = \sum_k p^k f^k(x^k, y)$$

with finally:

$$f^k(n,y) = a^k\Big(1 - \prod_0^{n-1} y'_m\Big) + \Big(\prod_0^{n-1} y'_m\Big)(y_n c^k + y'_n d^k) \text{ for } n \in \mathcal{N}$$

$$f^k(\infty,y) = a^k\Big(1 - \prod_0^{\infty} y'_m\Big) + \Big(\prod_0^{\infty} y'_m\Big) d^k$$

$$f^k(\partial,y) = a^k\Big(1 - \prod_0^{\infty} y'_m\Big) + \Big(\prod_0^{\infty} y'_m\Big) b^k$$

where $y = (y_0, \ldots, y_m, \ldots) \in Y$, y'_m denotes $1 - y_m$, and $\prod_0^{-1} = 1$.

Defining $\theta(y) = min\{(m; y_m = 1) \cup \{\infty\}\}$, it is clear that $\theta(y)$ determines the payoff, hence Y can also be written as $\overline{\mathcal{N}}$. Then one has, with $\xi \in X$:

$$f^k(\xi,m) = \mathrm{I}\{\xi \leq m-1\}d^k + \mathrm{I}\{\xi = m\}c^k + \mathrm{I}\{\xi > m\}a^k \text{ for } m \in \mathcal{N}$$
$$f^k(\xi,\infty) = \mathrm{I}\{\xi \leq \infty\}d^k + \mathrm{I}\{\xi = \partial\}b^k.$$

We write $\overline{\Gamma}(p)$ for the mixed extension of $\Gamma(p)$ where player 1's strategies are probabilities on (the borel subsets of) X^K (or as well K vector of probabilities on X, since f is decomposed on X^K), say χ in $\mathbf{X}^K$, and player 2's strategies are probabilities with finite support on Y, say Ψ in $\mathbf{Y}$.

Proposition 1: $\overline{\Gamma}(p)$ has a value $w(p)$.

Proof: For each $y, f^k(\cdot, y)$ is continuous on X. In fact either y corresponds to some m in $\mathcal{N}$ and f is constant on $x^k > m$, or to ∞ and f is constant on $\overline{\mathcal{N}}^K$. It follows that for (χ, Ψ) in $(\mathbf{X} \times \mathbf{Y})$:

$$F(\chi, \Psi) = \int f(x, y)\chi(dx)\Psi(dy)$$

is continuous in the first variable and affine in both. Moreover $\mathbf{X}^K$ is compact, hence by a corollary of Sion's theorem (see Mertens, Sorin and Zamir [5], Chapter 1), the game has a value (and player 1 has an optimal strategy).

4. The Result:

Theorem 2: $min\, max\, G(p)$ exists and $\overline{v}(p) = w(p)$.
The proof will follow from Lemmas 4 and 5.

Lemma 3: $\forall \Psi \in \mathbf{Y}$ there exists $z \in [0,1]^{\mathcal{N}}$ such that :

$$\int f(x, y)\Psi(dy) = f(x, z) \text{ for all } x \in X^K.$$

Proof: By the above remarks, Ψ can be described as the distribution of the stopping time θ. z_n is then just the conditional probability on the nth factor, given $\{\theta \geq n\}$.

Lemma 4: w satisfies (i).

Proof: Given $\varepsilon > 0$, let Ψ be an $\varepsilon/4$ optimal strategy of player 2 for $w(p)$, and by Lemma 3, represent Ψ by some z in $[0,1]^{\mathcal{N}}$.

Define now τ, strategy of player 2 in $G(p)$ as: play z_0 i.i.d. until the first Top of player 1, then play z_1 i.i.d. until the second Top, then z_2, and so on.

Let also $\rho = \prod_{m=0}^{\infty} z'_m$ and define N such that:

$$\text{if } \rho = 0, \qquad \prod_{m=0}^{N-1} z'_m < \varepsilon/4$$

$$\text{and if } \rho > 0, \qquad \prod_{m=N}^{\infty} z'_m > 1 - \varepsilon/4.$$

Let us now majorize $\overline{\gamma}_n(\sigma, \tau)$ for $n \geq N$. It is enough to consider each $\overline{\gamma}_n^k(\sigma^k, \tau)$. Since τ is independent of the previous moves of player 2, we can assume that σ^k

has the same property (one can replace at each stage σ^k by its expectation w.r.t. τ on J without changing the payoff). Moreover we can consider best replies and assume σ^k pure.

It follows then that σ^k is completely described by the dates $M_1, \ldots, M_m, \ldots$ of the successive Top, and define $M_0 = -1$.

We thus obtain, for $n \in (M_m, M_{m+1})$:

$$\begin{aligned}\gamma_n^k(\sigma,\tau) &= E_{\sigma^k,\tau}(g_n^k)\\ &= a^k\big(1-\prod_{\ell=0}^{m-1} z'_\ell\big) + \big(\prod_{\ell=0}^{m-1} z'_\ell\big)(z_m c^k + z'_m d^k)\\ &= f^k(m,z).\end{aligned}$$

Now for $n \geq M_N$, the expected stage payoff satisfies:
if $\rho = 0$, $|\gamma_n^k(\sigma,\tau) - a^k| \leq \varepsilon/2$, thus $|\gamma_n^k(\sigma,\tau) - f^k(\infty,z)| \leq \varepsilon/2$
if $\rho > 0$, either player 1 plays Bottom and one has:

$$|\gamma_n^k(\sigma,\tau) - (a^k(1-\rho) + \rho d^k)| \leq \varepsilon/2 \text{ by the choice of } N\text{, thus:}$$
$$|\gamma_n^k(\sigma,\tau) - f^k(\infty,z)| \leq \varepsilon/2,$$

or player 1 plays Top and we obtain similarly:

$$|\gamma_n^k(\sigma,\tau) - f^k(\partial,z)| \leq \varepsilon/2.$$

It follows that every expected stage payoff, except at most $K \times N$ of them, corresponding to $M_m, m = 1,\ldots,N$, for each σ^k, is within $\varepsilon/2$ of a feasible payoff against Ψ in $\overline{\Gamma}(p)$.

Hence $n \geq 8KN/\varepsilon$ implies:

$$\overline{\gamma}^p(\sigma,\tau) \leq w(p) + \varepsilon.$$

Lemma 5: w satisfies (ii).

Proof: Consider first χ, optimal strategy of player 1 in $\overline{\Gamma}(p)$. Given $\varepsilon > 0, \chi$ and τ strategy of player 2 in $G(p)$, we shall define σ, strategy of player 1 in $G(p)$ by the following procedure: we will introduce a family, indexed by X, of "non revealing" strategies in $G(p)$, i.e. transition probabilities from H to I, say α_x. σ^k will then be: select some α_x according to the probability χ^k on X. Let η be the stopping time of reaching the absorbing entry:

$$\eta = min\{(m; i_m = \text{Top}, j_m = \text{Left}) \cup \{\infty\}\}$$

and define N such that

$$\chi^k(N) \leq \varepsilon/3 \text{ for all } k \in K.$$

For each $m \leq N$, define inductively strategies α_m and times L_m as follows: α_0 is: always Bottom.

Given α_0 and τ, let $t_n^0 = \text{Prob}_{\alpha_0,\tau}(\jmath_n = \text{Left})$ and L_0 be the first time ℓ where:

$$\sum_k p^k \chi^k(0)(t_\ell^0 c^k + t_\ell^{0'} d^k) \leq \inf_n \left\{ \sum_k p^k \chi^k(0)(t_n^0 c^k + t_n^{0'} d^k) \right\} + \varepsilon/3.$$

Define now α_1 as: Bottom up to stage L_0 (excluded), Top at that stage L_0, and always Bottom after.

Similarly given α_m, let $t_n^m = \text{Prob}_{\alpha_m,\tau}(\jmath_n = \text{Left} \mid \eta > L_{m-1})$ and let L_m be the first $\ell > L_{m-1}$ where:

(1) $\sum_k p^k \chi^k(m)(t_\ell^m c^k + t_\ell^{m'} d^k) \leq \inf_n \left\{ \sum_k p^k \chi^k(m)(t_n^m c^k + t_n^{m'} d^k) \right\} + \varepsilon/3.$

α_{m+1} is then α_m up to stage L_m (excluded), Top at that stage and Bottom thereafter.

Now for $m > N$, we first introduce a new stopping time L' and a non revealing strategy α' satisfying :

(2) $\pi = \text{Prob}_{\alpha',\tau}(\eta \leq L') \geq \sup_{\alpha \in A} \text{Prob}_{\alpha,\tau}(\eta < \infty) - \varepsilon/9.$

where A is the set of strategies that coincide with α_N up to L_{N-1} (included).

If $N < m \leq \infty$, let $\alpha_m = \alpha_\infty$: play α' up to stage L' (included) and then Bottom for ever.

Finally we define α_∂ as: play α' up to stage L' (included) and then always Top.

Let us now introduce z in $[0,1]^{\bar{N}}$ satisfying: $z_m = t_{L_m}$, for $m < N$, $z_N = u$ where $\pi = 1 - \left(\prod_{m=0}^{N-1} y'_m\right)u'$ (note that $1 \geq \pi \geq 1 - \prod_{m=0}^{N-1} y'_m$), $z_m = 0$, for $m > N$, and we shall prove that for $n \geq L'$:

(3) $\gamma_n^p(\sigma,\tau) \geq \int f(x,z)\chi(dx) - 2\varepsilon/3.$

In fact we can decompose the above payoff on the events $\{\alpha_x \text{ is played }\}$, with x in X, so that:

$$\gamma_n^p(\sigma,\tau) = \sum_k p^k \gamma_n^k(\sigma^k,\tau) = \sum_k \sum_x p^k \text{ Prob}_{\sigma^k}(\alpha_x)\gamma_n^k(\alpha_x,\tau) = \sum_x \varphi_n(x,\tau).$$

For $m \le N$ we obtain:

$$\sum_k p^k \chi^k(m)\left\{\left(1-\prod_0^{m-1} y'_\ell\right)a^k + \prod_0^{m-1} y'_\ell(t_n^m c^k + t_n^{m'} d^k)\right\}$$

$$\ge \sum_k p^k \chi^k(m)\left\{\left(1-\prod_0^{m-1} y'_\ell\right)a^k + \prod_0^{m-1} y'_\ell(y_m c^k + y'_m d^k)\right\} - \varepsilon/3$$

$$\ge \sum_k p^k \chi^k(m) f^k(m,y) - \varepsilon/3.$$

For $N < m \le \infty$, we get:

$$\sum_k p^k \chi^k(m)\left\{\text{Prob}_{\alpha_m,\tau}(\eta \le n)a^k + (1-\text{Prob}_{\alpha_m,\tau}(\eta \le n))(t_n^m c^k + t_n^{m'} d^k)\right\}$$

$$\ge \sum_k p^k \chi^k(m)(\pi a^k + (1-\pi)d^k) - \varepsilon/3$$

since the choice of α' and L' implies $(1-\text{Prob}_{\alpha_m,\tau}(\eta \le n))t_n^m \le \varepsilon/9$.

Similarly, when α_∂ is used the payoff is at least:

$$\sum_k p^k \chi^k(\partial)(\pi a^k + (1-\pi)b^k) - \varepsilon/3.$$

It follows that for all $m \in X$, $m \ne N$, $\varphi_n(m,\tau) \ge \sum_k p^k \chi^k(m) f^k(m,z) - \varepsilon/3$. Since, by the choice of N, $|\varphi_n(N,\tau)|$ as well as $|\sum_k p^k \chi^k(N) f^k(N,z)|$ are bounded by $\varepsilon/3$, we obtain (3) by summing.

Hence $n \ge 6L'/\varepsilon$ implies:

$$\overline{\gamma}_n^p(\sigma,\tau) \ge w(p) - \varepsilon.$$

This achieves the proof of Theorem 2.

5. A Geometric Approach: We shall here obtain the minmax by describing its supporting hyperplanes; equivalently, we will construct a convex set S in $\mathcal{R}^K$ with: $\overline{v}(p) = min_{s \in S}\langle p, s\rangle$.

5.1. General results: (see Sorin, [8], p.187)

We first need some definitions:

Player 1 can *force a set* S in $\mathcal{R}^K$, if for every $\varepsilon > 0$ and every τ there exists a strategy σ and a number N, such that $n \geq N$ implies that $\overline{\gamma}_n(\sigma,\tau) = \{\overline{\gamma}_n^k(\sigma,\tau)\}_{k\in K}$ belongs to S_ε^+, with $S_\varepsilon^+ = \{t \in \mathcal{R}^K;$ there exists $s \in S, t^k \geq s^k - \varepsilon, \forall k \in K\}$. (Write S^+ for S_0^+).

Player 2 can *guarantee a point* M in $\mathcal{R}^K$, if for every $\varepsilon > 0$ there exists a strategy τ and a number N, such that for every σ, $n \geq N$ implies: $\overline{\gamma}_n^k(\sigma,\tau) \leq M^k + \varepsilon$, for all k.

It is easy to see that the set S_2 of points that player 2 can guarantee is convex. On the other hand if $H(p,u)$ denotes the half space $\{t \in \mathcal{R}^K; \langle p,t\rangle \geq u\}$ (with p probability on K) and player 1 can force $H(p,g(p))$ for all p, then player 1 can force $H(p, \mathrm{Cav}\, g(p))$. Let then S_1 be the intersection of all half spaces $H(p,u)$ that player 1 can force. We now obtain:

Proposition 6: $\overline{v}(p)$ exists and equals $min_{s\in S}\langle p,s\rangle \iff S_1 = S_2 (= S)$ (S is then called the minmax set).

5.2. Description of S:

We assume from now on $\#K = 2$.

We denote by A (resp. B, C, D) the point (a^1,a^2) and define $V = (v^1,v^2)$ where v^k is the value (minmax would suffice) of $G(k)$. We also introduce $Q = (a^1,c^2), P = (v^1,d^2)$ and similarly $Q' = (a^2,c^1), P' = (d^1,v^2)$. Given two points M and N, let $M \vee N = (max\{m^1,n^1\}, max\{m^2,n^2\})$ and finally $[M,N]$ is the line segment MN.

Lemma 7: Player 1 can force $\{V\}$ (I), $[C,D]$ (II), $[A, B \vee D]$ (III), $[Q,P]$ (IV) and $[Q',P']$ (V).

Proof: (I) and (II) are clear. For (III), player 1 uses first a strategy that achieves the maximal probability of getting the absorbing payoff and plays then either Top or Bottom for ever (cf. (2) in Lemma 5). As for (IV), player 1 plays always Bottom in game 2 and Bottom in game 1 until the stage where the expected (non absorbing) payoff in game 2 is minimal. He plays then Top (in game 1) and thereafter optimally. The payoff in game 1 will then be some $ta^1 + (1-t)v^1$ in game 1 and in game 2 some average of expressions like $t_n c^2 + (1-t_n)d^2 \geq tc^2 + (1-t)d^2$, (cf. (1) in Lemma 5).

Lemma 8: Player 2 can guarantee $(\alpha) B \vee D, (\beta)(xC + x'D) \vee A$ for all $x \in [0,1]$. Moreover, (γ) if player 2 can guarantee M he can also guarantee $(xC + x'D) \vee (xA + x'M)$.

Proof: (α) and (β) are clear. As for (γ), player 2 plays $(x, 1-x)$, i.i.d. until player 1 plays Top. He then approaches M.

These strategies are sufficient to obtain the following result:

Proposition 9: Let S be the intersection of the half spaces $H(p,u)$ containing one of the sets T, $T = I, \ldots, V$. Then S is the minmax set.

Proof: It is enough to prove that the extreme points of the frontier of S belong to S_2.

1) If $D \in A^+$, player 2 can approach $B \vee D(\alpha)$ and $A(\beta)$; hence $S = [A, B \vee D]^+$.
2) If $A \in D^+$, either $C \in D^+$ or $D \in C^+$ and $S = V^+$; or suppose $c^1 > d^1$ and let x satisfy $xc^2 + (1-x)d^2 = v^2$. Then $E = xC + x'D$ belongs to S_2 and $S = [D, E]^+$.
3) In the remaining case we can assume $a^1 > d^1$ and $a^2 < d^2$.

The analysis is done by considering the different values of B and C and can be reduced to the 2 following configurations :

First case:

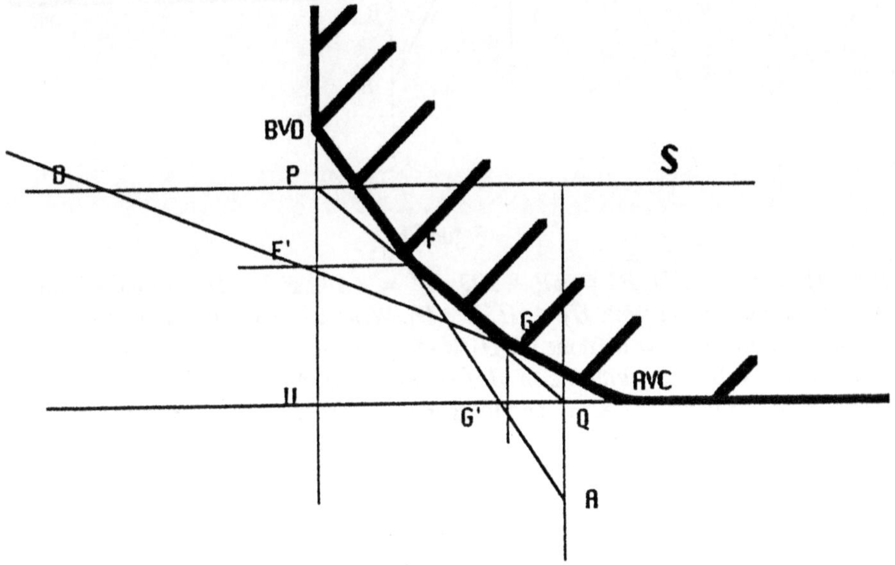

Figure 1

Player 2 can approach $B \vee D(\alpha)$ and $C \vee A(\beta)$.

For F, write $F = xA + x'B \vee D = xQ + x'P$, hence $F' = xC + x'D$ so that using $M = B \vee D$ in (γ), $F \in S_2$.

Similarly for G, write $G = yC + y'D = yQ + y'P$, so that $G' = yA + y'B \vee D$ and as above $G \in S_2$.

Second case:

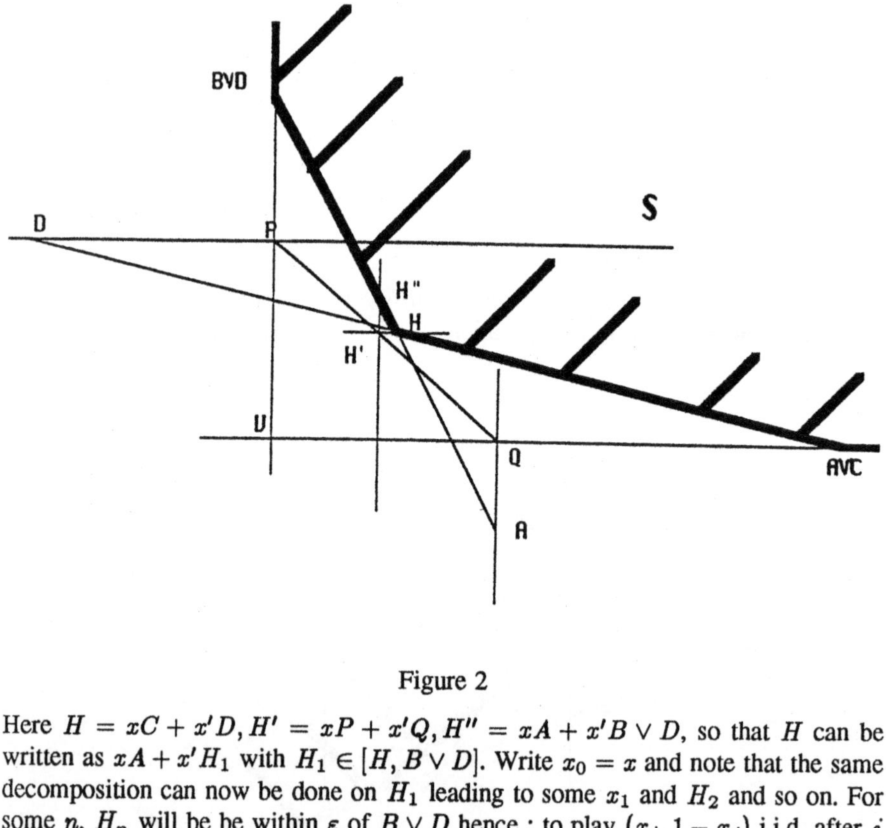

Figure 2

Here $H = xC + x'D, H' = xP + x'Q, H'' = xA + x'B \vee D$, so that H can be written as $xA + x'H_1$ with $H_1 \in [H, B \vee D]$. Write $x_0 = x$ and note that the same decomposition can now be done on H_1 leading to some x_1 and H_2 and so on. For some n, H_n will be be within ε of $B \vee D$ hence : to play $(x_j, 1 - x_j)$ i.i.d. after j moves Top of player 1 ($j < n$), and to guarantee $B \vee D$ after n Top is the required strategy.

Remarks: The above analysis indicates that for $\#K = 2$, player 1 can be restricted to use strategies with support on $\{\alpha_x\}, x = 0, 1, \infty, \partial$. Considering then only best replies of player 2, it follows that $\overline{v}(p)$ is the value of a (one-shot) matrix game with lack of information on one side (hence, as in Proposition 9, piece-wise linear).

On the other hand it is easy to see that one cannot restrict player 2 to use strategies in $\overline{\Gamma}(p)$ with θ uniformly bounded or ∞. In other words, given $\varepsilon > 0$, the number $N(\varepsilon)$ of "exceptional moves" (like in the proof of Proposition 9, second case) needed before using some i.i.d. strategy cannot be uniformly bounded.

A very similar, but simpler structure, is obtained when only one entry is non absorbing. When the absorbing states are on the diagonal, the minmax is again the value of some auxiliary game in normal form with lack of information on one side (Sorin, [9]).

Acknowledgements: It is a pleasure to thank J.F. Mertens for nice comments.

References

[1] *Blackwell, D.* (1956). An Analog of the Minimax Theorem for Vector Payoffs *Pacific Journal of Mathematics*, **6**, 1-8.

[2] *Blackwell, D. and T. S. Ferguson* (1968). The Big Match, *Annals of Mathematical Statistics*, **39**, 159-163.

[3] *Melolidakis, C.* (1984). Ph.D. Thesis, UCLA.

[4] *Mertens, J.F.* (1987). Repeated Games, *Proceeding of the International Congress of Mathematicians, Berkeley* (1986), 1528-1577.

[5] *Mertens, J.F., S. Sorin and S. Zamir* (1987). "Repeated Games", book to appear.

[6] *Mertens, J.F. and S. Zamir* (1976). On a Repeated Game without a Recursive Structure, *International Journal of Game Theory*, **5**, 173-182.

[7] *Sorin, S.* (1984). "Big match" with Lack of Information on One Side (Part I), *International Journal of Game Theory*, **13**, 201-255.

[8] *Sorin, S.* (1985). "Big match" with Lack of Information on One Side (Part II), *International Journal of Game Theory*, **14**, 173-204.

[9] *Sorin, S.* (1987). More on the 'Big match' with Lack of Information on One Side, manuscript.

[10] *Waternaux, C.* (1983). Solution for a Class of Games without Recursive Structure, *International Journal of Game Theory*, **12**, 129-160.

[11] *Waternaux, C.* (1983). Minmax and Maxmin of Repeated Games without a Recursive Structure, Core D.P. 8313.

Sylvain Sorin
Mathematiques
Université Louis Pasteur
7 Rue René Descartes
67084 Strassbourg
France

Shmuel Zamir
Department of Economics
University of Pittsburgh
Pittsburgh, PA 15260
U.S.A.

Stochastic Games with Lack of Information on One Side and Positive Stop Probabilities

Costis Melolidakis

Abstract: We examine Stochastic Games with Lack of Information on One Side (SGLIOS) under the assumption that at each state the game ends with positive probability. The existence of a myopic, stationary in some sense, optimal strategy for the informed player is established. The main tool to approach the problem of control under incomplete information is an updating mechanism using posterior probabilities on the state space given the history up to the present stage.

1. Introduction: In 1953 L. S. Shapley introduced Stochastic Games with positive stop probabilities [1] and showed that their value exists and that both players possess optimal stationary strategies. His results opened the area of stochastic games and were generalized in many directions as it is well known.

In this paper, we keep the original Shapley's structure unaltered except one informational complication: The second player (minimizer) is never informed of the true state of nature, i.e. he does not know the actual matrix game played at each stage, but he is informed of the end of the game. We call such games "Stochastic Games with Lack of Information on One Side (SGLIOS) with positive stop probabilities".

One way to deal with the informational difficulty in such games is through the use of an updating mechanism that uses posterior probabilities on the state space given the history up to the present stage. It is often the case that the informed player may restrict his attention to strategies that depend on the history only through these conditional probabilities.

This general methodology for the "reduction" of the problem is not new: Among other cases, it has been used in Dynamic Programming situations that involve "learning" (e.g. the "Two Armed Bandit Problem", see [2] or [3]) and in Dynamic

Game Theory situations with incomplete information (e.g. in "Repeated Games With Lack of Information on One Side", see [4]).

In the general context of SGLIOS, not necessarily with positive stop probabilities, this approach leads to an interesting result: The stochastic game with lack of information on one side generates an ordinary and intuitively reasonable, but continuous, stochastic game with the following property: If both games have a value, then the values are the same, under any usual definition of the payoff. This result was obtained by C. Melolidakis in 1984, see [5]. In 1986, J.F. Mertens, using the same approach, obtained pertinent results in a more general framework [11].

If one can guarantee the value in both the original and the generated stochastic game, then, he may use properties of the strategies for the informed player in the generated stochastic game to describe his strategies in the original game. In the case of discounted SGLIOS, this was done in [5] (see also [6]).

The proof (or disproof) of the stronger statement: If the original game has a value, then the generated stochastic game has a value (and, of course, they are the same) is, as far as I know, an open problem.

In the case of SGLIOS with positive stop probabilities, the above central result is sufficient to establish certain properties for the value of the game and the optimal strategies of the informed player. This is due to the fact that such a game ends after a finite number of steps with probability one.

SGLIOS with "zero stop probabilities" are much more difficult to handle and few things are known in this area. In particular, S. Sorin [12,13] has studied cases with absorbing payoffs and C. Melolidakis [5,9] cases where the transitions do not depend on the actions of the two players.

2. Formulation of the Problem:

Definition 1: By SGLIOS with positive stop probabilities we mean a 0-sum 2 person game defined by :

i) A set $S = \{G^1, G^2, \ldots, G^k\}$ of $m \times n$ matrices. They are the "states". We adjoin $\emptyset$ to S, where $\emptyset$ represents the end of the game.

ii) An initial probability distribution $P_1 = (p_{11}, \ldots, p_{1k})$ on S.

iii) Transition probabilities $q(\cdot \mid r, i, j)$. They are distributions on $S \cup \{\emptyset\}$ which depend on the integers r, i, j with $1 \leq r \leq k$, $1 \leq i \leq m$, $1 \leq j \leq n$. We assume that $\sum_{r' \in S} q(r' \mid r, i, j) < 1 \; \forall \, r, i, j$.

Let us denote this game by Γ. The interpretation is the following: Γ is played in stages. First nature chooses an initial state r according to p_1 and informs player I but not II about her choice. Then, the two players make their moves i and j simultaneously, they are both informed of their common choices (i, j) and the game either ends with stop probability $q(\emptyset \mid r, i, j) = 1 - \sum_{r' \in S} q(r' \mid r, i, j) > 0$ or the system jumps to a new state r' according to $q(\cdot \mid r, i, j)$. I is again informed about the true state r' but II isn't, and the process continues indefinitely. When the game ends, both players are informed of this.

At each stage the immediate payoff G^r_{ij} enters into a list kept by nature. Although the players remember their moves i and j, they are not informed of the G^r_{ij} (which of course I independently knows).

So, provided the game has not ended, their common knowledge before their moves at stage N is the string $(i_1, j_1, \ldots, i_{N-1}, j_{N-1})$ which we call *history* and denote by h_N. Player I also knows the string of passed and present states $(\theta_1, \theta_2, \ldots, \theta_N)$.

Let $\mathrm{H}_N = [\{1,\ldots,m\} \times \{1,\ldots,n\}]^{N-1}$ and $\mathcal{A}_N = \{1,\ldots,k\}^N$. Since Γ is a game of perfect recall both players may restrict their attention to behavioral strategies (see [7]). So from now on "strategy"means behavioral strategy. A strategy for I is a sequence $\sigma = (\sigma_1, \sigma_2, \ldots, \sigma_N, \ldots)$ such that σ_N is an $\mathrm{H}_N \times \mathcal{A}_{N-1}$ measurable function consisting of k probability distributions over $\{1,\ldots,m\}$ - one for each possible present state - i.e. $\sigma_N = (s_1^N, s_2^N, \ldots, s_k^N) \in (P^m)^k$, where P^m denotes the probability simplex in $\mathbf{R}^m$. Similarly a strategy for II is a sequence $\tau = (\tau_1, \ldots, \tau_N, \ldots)$ such that τ_N is an H_N measurable function consisting of one probability distribution over $\{1,\ldots,N\}$.

Now, let T be a random variable (r.v.) representing the stage the game ends, I_N, J_N be r.v.'s representing the actions of I and II respectively at stage N, and H_N a r.v. representing the history up to stage N. Then, σ, τ, p_1, and q induce a probability measure P_N on $\mathcal{A}_N \times \mathrm{H}_N$ which can be uniquely extended to a measure P on $\mathcal{A}_\infty \times \mathrm{H}_\infty$ (Kolmogorov's consistency theorem). Then the appropriate marginals of this extension will give the distributions of H_N, I_N, J_N and Θ_N $\forall\, N \in \mathbf{N}$, ($\Theta_N$ is a r.v. representing the state).

Let us also denote by $A_N(\sigma,\tau)$ the expected immediate payoff at stage N, provided that the game has not ended, i.e.

$$A_N(\sigma,\tau) := E_{\sigma,\tau}(G^{\Theta_N}_{I_N J_N} \mid T > N).$$

Denote also $A_N(\sigma,\tau)P(T > N)$ by $R_N(\sigma,\tau)$.

The total payoff $H(\sigma,\tau)$ in Γ is then given by

$$H(\sigma,\tau) = \sum_{N=1}^{\infty} R_N(\sigma,\tau). \tag{1}$$

Notice that, as in the original paper by Shapley, by letting $1 - \beta = min_{r,i,j}\, q(\emptyset \mid r, i, j)$ and $M = max_{r,i,j} |G^r_{ij}|$ we may see that (1) is absolutely dominated term by term by $\beta M + \beta^2 M + \beta^3 M + \ldots = \beta M/(1-\beta)$ and the payoff is therefore well defined, since $1 - \beta > 0$ from the assumptions.

The next part is a modification of a more general result applying to all Stochastic Games with Lack of Information on One Side, which first appeared in [5] (chapter 2, see also [8]). Here we give a detailed proof for reasons of completeness.

3. The Generated Stochastic Game:

Proposition 1: In Γ, player I may forget the previous string of states, i.e., $\forall\sigma$ of I $\exists\sigma^*$ of I such that: σ^*_N is H_N measurable only and $R_N(\sigma^*, \tau) = R_N(\sigma, \tau)$ $\forall N \in \mathbf{N}$ with τ any strategy of II.

Proof: Let P denote the probability measure $P_{\sigma,\tau}$, induced on $\mathcal{A}_\infty \times H_\infty$ by σ, τ, p_1 and q. Define σ^* by defining:

$$\sigma^*_N(i_N) := (s^*_1(i_N), s^*_2(i_N), \ldots, s^*_k(i_N)),$$

where $s^*_{\theta_N}(i_N) = P_{\sigma,\tau}(I_N = i_N \mid \Theta_N = \theta_N, H_N = h_N, T > N)$. Then define P^* to be the probability measure $P_{\sigma^*,\tau}$, induced on $\mathcal{A}_\infty \times H_\infty$ by σ^*, τ, p_1 and q. Then, to prove the proposition it is sufficient to show that

$$P(\Theta_N = \theta_N, I_N = i_N, J_N = j_N \mid T > N) = P^*(\Theta_N = \theta_N, I_N = i_N, J_N = j_N \mid T > N) \tag{2}$$

and that,

$$P(T > N) = P^*(T > N), \tag{3}$$

since,

$$R_N(\sigma, \tau) = \sum_{(\theta_N, i_N, j_N)} P(\Theta_N = \theta_N, I_N = i_N, J_N = j_N \mid T > N) G^{\theta_N}_{i_N j_N} P(T > N).$$

The proof of (2) may be done by following exactly the same steps as in the corresponding proposition of [5] or [9], after noticing that

$$P(X_N = x_N \mid Y_N = y_N, T > N) = P(X_N = x_N \mid Y_N = y_N, T > N + C)$$

for $C > 0$ and X_N, Y_N any pair of random variables chosen from $\{\Theta_N, I_N, J_N\}$. The proof of (3) is by induction on N and is omitted. ◇

So, from now on, we may restrict attention to strategies of I that do not depend on the previous string of states without any loss of generality.

Proposition 2: Γ may be played in the following way: The players choose σ and τ and give it to "nature" who then at stage 1 chooses θ_1 according to p_1. Then, i_1 and j_1 are chosen according to σ_1 and τ_1 and $G^{\theta_1}_{i_1 j_1}$ enters a list kept by nature. Then, "nature" decides whether the game is going to end according to $P(T = 2 \mid I_1 = i_1, J_1 = j_1)$. Provided the game continues, nature chooses θ_2 according to $p_2 := P(\Theta_2 = \theta_2 \mid I_1 = i_1, J_1 = j_1, T > 2)$ and the process continues. So at stage N, nature decides whether the game will end according to $P(T = N \mid T > N-1, H_N = h_N)$ and, provided the game continues, θ_N is chosen according to $p_N := \text{Prob}(\Theta_N = \theta_N \mid H_N = h_N, T > N)$. Then i_N and j_N are chosen according to σ_N - using this θ_N - and τ_N. The immediate payoff $G^{\theta_N}_{i_N j_N}$ enters nature's list who then computes p_{N+1} using h_N and θ_N, i_N and j_N. Then, the above procedure is repeated, i.e. nature chooses θ_{N+1} according to p_{N+1} etc.

Proof: The proof is by establishing the following relation,

$$R_N(\sigma,\tau) = \sum_{h_N} \Big\{ p(H_N = h_N) \cdot \sum_{\theta_N, i_N, j_N} [\sigma(i_N \mid \theta_N, h_N, T > N) \cdot \tau(j_N \mid h_N, T > N) \cdot P(\theta_N \mid h_N, T > N) \cdot G^{\theta_N}_{i_N j_N}] \cdot P(T > N \mid H_N = h_N) \Big\}.$$

This may be done by techniques similar to those used in proving Proposition 1. See also [5] or [9]. ◇

Remark: Computations give the formulas for the stop-probabilities at stage N and for $p_N := \text{Prob}(\Theta_N = \theta_N \mid H_N = h_N, T > N)$. We have

$$\text{Prob}(T = N \mid H_N = h_N, T > N-1) = \frac{\sum_{\ell \in S} q(\emptyset \mid \ell, i_{N-1}, j_{N-1}) s_\ell(i_{N-1})(p_{N-1})_\ell}{\sum_{\ell \in S} s_\ell(i_{N-1})(p_{N-1})_\ell} \tag{4}$$

Notice also that $\text{Prob}(T = N \mid H_N = h_N, T > N-1) \geq 1 - \beta > 0$.

$$\begin{aligned}(p_N)_r &= \frac{1}{1 - \text{Prob}(T = N \mid H_N = h_n, T > N-1)} \cdot \frac{\sum_{\ell \in S} q(r \mid \ell, i_{N-1}, j_{N-1}) s_\ell(i_{N-1})(p_{N-1})_\ell}{\sum_{\ell \in S} s_\ell(i_{N-1})(p_{N-1})_\ell} \\ &= \frac{\sum_{\ell \in S} q(r \mid \ell, i_{N-1}, j_{N-1}) s_\ell(i_{N-1})(p_{N-1})_\ell}{\sum_{r \in S}\sum_{\ell \in S} q(r \mid \ell, i_{N-1}, j_{N-1}) s_\ell(i_{N-1})(p_{N-1})_\ell}\end{aligned} \tag{5}$$

At this point we have to introduce some additional notation. Let $j^N = (j_1, j_2, \ldots, j_N)$, $i^N = (i_1, i_2, \ldots, i_N)$, $\sigma^N = (\sigma_1, \sigma_2, \ldots, \sigma_N)$, $p^N = (p_1, p_2, \ldots, p_N)$.

In Γ, player I takes notice of i^{N-1}, j^{N-1} and then, provided the game continues, he chooses an action $i_N \in \{1, 2, \ldots, m\}$ according to $\sigma_N = (s_1^N, s_2^N, \ldots, s_k^N)$. Similarly, II notices i^{N-1}, j^{N-1} and then, provided the game continues, she chooses an action j_N according to τ_N.

Let us now define an *extension* of Γ, $\hat{\Gamma}$ in the following way. Provided that the game does not end at stage N, player I takes notice of i^{N-1}, j^{N-1}, σ^{N-1} and p^N and then chooses $\sigma_N = (s_1^N, \ldots, s_k^N) \in (P^m)^k$ according to $\hat{\sigma}_N = (\hat{s}_1^N, \ldots, \hat{s}_k^N)$

with $\hat{s}_r^N$, $r = 1, \ldots, k$ a distribution on the Borel sets of P^m. Again, provided the game continues, player II notices i^{N-1}, j^{N-1}, σ^{N-1} and p^N and then chooses j_N according to $\hat{\tau}_N$, a probability distribution on $\{1, \ldots, n\}$. In $\widehat{\Gamma}$, states are chosen by the mechanism described in Proposition 2.

As before $\hat{\sigma} := (\hat{\sigma}_1, \hat{\sigma}_2, \ldots \hat{\sigma}_N, \ldots)$ and $\hat{\tau} = (\hat{\tau}_1, \ldots, \hat{\tau}_N, \ldots)$ determine a probability measure $\widehat{P}$ on $\mathcal{A}_\infty \times I_\infty \times J_\infty \times \Sigma_\infty \times P_\infty$, the space of sequences of θ's, i's , j's , σ's and p's.

The expected immediate payoff is defined by in the same way as in Γ i.e. $\widehat{A}_N(\sigma, \tau) := E_{\hat{\sigma},\hat{\tau}}(G^{\Theta_N}_{I_N J_N} \mid T > N)$, where now expectations are taken with respect to $\widehat{P}$, e.g.
$\widehat{P}(I_N = i_N \mid C) = (E_{\hat{\sigma},\hat{\tau}}(\sigma_N \mid C))(i_N)$, where C is some conditioning. We then get:

Lemma 1: If Γ has a value then $\widehat{\Gamma}$ has a value and $\mathrm{val}(\widehat{\Gamma}) = \mathrm{val}(\Gamma)$.

Proof: If $\mathrm{val}(\Gamma)$ exists then I may ignore the additional information while playing $\widehat{\Gamma}$ and use as $\hat{s}^N_{\theta_N}$ the probability distribution that gives all its weight on $s^N_{\theta_N}$ with $s^N_{\theta_N}$ an ϵ-optimal strategy in Γ. Then I guarantees $\mathrm{val}(\Gamma)$ in $\widehat{\Gamma}$ since the additional information of $\hat{\sigma}_N$ and p_N cannot help player II . The argument is similar for player II. ◇

Now define a *restriction* $\widehat{\Gamma}^*$ of $\widehat{\Gamma}$ by restricting the strategy spaces of the two players. The strategies of I and II in $\widehat{\Gamma}^*$ depend on j^{N-1}, σ^{N-1} and p^N but not in i^{N-1}. Let us denote the strategies of $\widehat{\Gamma}^*$ viewed as strategies in $\widehat{\Gamma}$ by (*).

Lemma 2: If any one of the two players restricts his attention to (*) strategies while playing $\widehat{\Gamma}$, then, the other can do as well by also restricting his attention to (*) strategies, while playing $\widehat{\Gamma}$.

Proof: Let $\hat{\tau}$ be of the form $\hat{\tau}_N = P(J_N = j_N \mid j^{N-1}, \sigma^{N-1}, p^N, T > N)$, i.e. $\hat{\tau}$ belongs in the subspace of (*) strategies. We will show that I may as well restrict his $\hat{\sigma}$ strategy, which in general will depend at stage N on θ_N, i^{N-1}, j^{N-1}, σ^{N-1} and p^N in the (*) space, i.e. he may ignore i^{N-1} then.

Define $\hat{\sigma}^*_N := (\hat{s}_1^{*N}, \ldots, \hat{s}_k^{*N})$ with $\hat{s}^{*N}_{\theta_N} = P_{\hat{\sigma}\hat{\tau}}(\sigma_N \mid \theta_N, j^{N-1}, \sigma^{N-1}, p^N, T > N)$. Then, a new probability measure is induced by $(\hat{\sigma}^*, \hat{\tau})$ on $\mathcal{A}_\infty \times I_\infty \times J_\infty \times \Sigma_\infty \times P_\infty$, the $P_{\hat{\sigma}^*\hat{\tau}}$. Let us denote $P_{\hat{\sigma}\hat{\tau}}$ by P and $P_{\hat{\sigma}^*\hat{\tau}}$ by P^*. We will show that,

$$P(\theta_N, i_N, j_N \mid T > N) = P^*(\theta_N, i_N, j_N \mid T > N) \tag{6}$$

and that

$$P(T > N) = P^*(T > N). \tag{7}$$

We first get,

$$P(\theta_N, i_N, j_N \mid T > N) = \sum_{j^{N-1},\sigma^{N-1},p^N} P(i_N \mid \theta_N, j^{N-1}, \sigma^{N-1}, p^N, j_N, T > N)\cdot$$
$$P(\theta_N, j_N \mid j^{N-1}, \sigma^{N-1}, p^N, T > N) P(j^{N-1}, \sigma^{N-1}, p^N \mid T > N) \quad (8)$$

Given $j^{N-1}, \sigma^{N-1}, p^N$ and $T > N$, the distribution of j_N is independent of i_N. Hence,

$$\begin{aligned} P(i_N \mid \theta_N, j^{N-1}, \sigma^{N-1}, p^N, j_N, T > N) &= P(i_N \mid \theta_N, j^{N-1}, \sigma^{N-1}, p^N, T > N) \\ &= (E_{\hat{\sigma}\hat{\tau}}[\sigma_N \mid j^{N-1}, \sigma^{N-1}, p^N, T > N])_{\theta_N}(i_N) \\ &= E_{\hat{\sigma}^*\hat{\tau}}[\sigma_N]_{\theta_N}(i_N) \\ &= P^*(i_N \mid \theta_N, j^{N-1}, \sigma^{N-1}, p^N, T > N). \end{aligned} \quad (9)$$

Also,

$$P(\theta_N, j_N \mid j^{N-1}, \sigma^{N-1}, p^N, T > N) = p_N(\theta_N) \cdot P(j_N \mid j^{N-1}, \sigma^{N-1}, p^N, T > N). \quad (10)$$

Since τ is the same in P and P^* and since p_N is given we conclude that,

$$P(\theta_N, j_N \mid j^{N-1}, \sigma^{N-1}, p^N, T > N) = P^*(\theta_N, j_N \mid j^{N-1}, \sigma^{N-1}, p^N, T > N). \quad (11)$$

On the other hand, certainly,

$$P(j^{N-1}, \sigma^{N-1}, p^N \mid T > N) = P(j_{N-1}, \sigma_{N-1}, p_N \mid j^{N-2}, \sigma^{N-2}, p^{N-1}, T > N)\cdot$$
$$P(j^{N-2}, \sigma^{N-2}, p^{N-1}, T > N-1).$$

The result will follow by induction if we show that,

$$P(j_{N-1}, \sigma_{N-1}, p_N \mid j^{N-2}, \sigma^{N-2}, p^{N-1}, T > N) = P^*(j_{N-1}, \sigma_{N-1}, p_N \mid j^{N-2}, \sigma^{N-2}, p^{N-1}, T > N).$$

But,

$$P(j_{N-1}, \sigma_{N-1}, p_N \mid j^{N-2}, \sigma^{N-2}, p^{N-1}, T > N) = P(\sigma_{N-1}, p_N \mid j^{N-1}, \sigma^{N-2}, p^{N-1}, T > N)\cdot$$
$$P(j_{N-1} \mid j^{N-2}, \sigma^{N-2}, p^{N-1}, T > N-1).$$

Since the last one is equal to $P^*(j_{N-1} \mid j^{N-2}, \sigma^{N-2}, p^{N-1}, T > N-1)$ - τ is in (*) - it remains to show that,

$$P(\sigma_{N-1}, p_N \mid j^{N-1}, \sigma^{N-2}, p^{N-1}, T > N) = P^*(\sigma_{N-1}, p_N \mid j^{N-1}, \sigma^{N-2}, p^{N-1}, T > N). \quad (12)$$

Now,

$$P(\sigma_{N-1}, p_N \mid j^{N-1}, \sigma^{N-2}, p^{N-1}, T > N) = P(p_N \mid j^{N-1}, \sigma^{N-1}, p^{N-1}, T > N) \cdot P(\sigma_{N-1} \mid j^{N-1}, \sigma^{N-2}, p^{N-1}, T > N-1). \quad (13)$$

and

$$P(p_N \mid j^{N-1}, \sigma^{N-1}, p^{N-1}, T > N) = \sum_{i_{N-1}, \theta_{N-1}} [P(p_N \mid \theta_{N-1}, i_{N-1}, j^{N-1}, \sigma^{N-1}, p^{N-1}, T > N) \cdot P(\theta_{N-1} \mid j^{N-1}, \sigma^{N-1}, p^{N-1}, T > N-1) \cdot P(i_{N-1} \mid \theta_{N-1}, j^{N-1}, \sigma^{N-1}, p^{N-1}, T > N-1)].$$

But,

$$P(\theta_{N-1} \mid j^{N-1}, \sigma^{N-1}, p^{N-1}, T > N-1) \cdot P(i_{N-1} \mid \theta_{N-1}, j^{N-1}, \sigma^{N-1}, p^{N-1}, T > N-1) = p_{N-1}(\theta_{N-1})(\sigma_{N-1})_{\theta_{N-1}}(i_{N-1}).$$

And, $P(p_N \mid \theta_{N-1}, i_{N-1}, j^{N-1}, \sigma^{N-1}, p^{N-1}, T > N) = 0$ or 1 according to whether i_{N-1} is compatible with $\sigma_{N-1}, j_{N-1}, p_{N-1}, p_N$. Hence, since p_{N-1} and σ_{N-1} are given, we conclude that:

$$P(p_N \mid j^{N-1}, \sigma^{N-1}, p^{N-1}, T > N) = P^*(p_N \mid j^{N-1}, \sigma^{N-1}, p^{N-1}, T > N). \quad (14)$$

Also,

$$P(\sigma_{N-1} \mid j^{N-1}, \sigma^{N-2}, p^{N-1}, T > N-1) = \sum_{\theta_{N-1}} P(\sigma_{N-1} \mid \theta_{N-1}, j^{N-1}, \sigma^{N-2}, p^{N-1}, T > N-1) \cdot P(\theta_{N-1} \mid j^{N-1}, \sigma^{N-2}, p^{N-1}, T > N-1).$$

Now since given $\theta_{N-1}, j^{N-2}, \sigma^{N-2}, p^{N-1}$ and $T > N-1$ the distribution of σ_{N-1} is independent of j_{N-1}, and since - from the definition of P^*-

$$P(\sigma_{N-1} \mid \theta_{N-1}, j^{N-2}, \sigma^{N-2}, p^{N-1}, T > N-1) = \\ P^*(\sigma_{N-1} \mid \theta_{N-1}, j^{N-2}, \sigma^{N-2}, p^{N-1}, T > N-1)$$

we conclude that,

$$P(\sigma_{N-1} \mid j^{N-1}, \sigma^{N-2}, p^{N-1}, T > N-1) = \\ \sum_{\theta_{N-1}} P^*(\sigma_{N-1} \mid \theta_{N-1}, j^{N-2}, \sigma^{N-2}, p^{N-1}, T > N-1) p_{N-1}(\theta_{N-1})$$

which leads to:

$$P(\sigma_{N-1} \mid j^{N-1}, \sigma^{N-2}, p^{N-1}, T > N-1) = \\ P^*(\sigma_{N-1} \mid j^{N-1}, \sigma^{N-2}, p^{N-1}, T > N-1). \quad (15)$$

Then (13), (14) and (15) show that (12) is true, which leads to

$$P(j^{N-1}, \sigma^{N-1}, p^N \mid T > N) = P^*(j^{N-1}, \sigma^{N-1}, p^N \mid T > N).$$

Now (8), (9) and (11) establish (6) , i.e.

$$P(\theta_N, i_N, j_N \mid T > N) = P^*(\theta_N, i_N, j_N \mid T > N).$$

It remains to establish (7). But,

$$P(T > N) = P(T > N \mid T > N-1) P(T > N-1)$$

and, using similar to the previous arguments, one can show that

$$P(T > N \mid T > N-1) = P^*(T > N \mid T > N-1).$$

Hence (7) follows by induction. Finally, (6) and (7) lead to

$$R_N(\hat\sigma, \hat\tau) = R_N(\hat\sigma^*, \hat\tau),$$

where $R_N(\hat\sigma, \hat\tau)$ denotes the expected immediate payoff of $\hat\Gamma$ at stage N. Hence, the first half of the lemma has been proved, provided that $\hat\sigma^*$ is independent of $\hat\tau$. But this last statement is true, since:

$$P_{\hat\sigma,\hat\tau}(\sigma_N \mid \theta_N, j^{N-1}, \sigma^{N-1}, p^N, T > N) = \frac{\sum_{\theta^{N-1}} P(\sigma^N, j^{N-1}, \theta^N, p^N \mid T > N)}{\sum_{\theta^{N-1}} P(\sigma^{N-1}, j^{N-1}, \theta^N, p^N \mid T > N)}$$

and if we write down numerator and denominator explicitly we see that this probability doesn't depend on τ, since τ is independent of θ 's and hence factors out and cancels.

The corresponding result for II is also true, i.e. if I keeps his strategy $\hat\sigma$ in the subspace (*), II may as well restrict his own strategy $\hat\tau$ in (*). The proof follows exactly the same steps as before, where now

$$\hat\tau_N^* = P_{\hat\sigma\hat\tau}(J_N = j_N \mid j^{N-1}, \sigma^{N-1}, p^N, T > N)$$

while the independence of $\hat\tau^*$ and $\hat\sigma$ is checked by noting that given j^{N-1}, σ^{N-1}, p^N and $T > N$, $\hat\tau^*$ doesn't depend on $\hat\sigma$, since only $\sigma_1, \ldots, \sigma_{N-1}$ appear in its formula. ◇

Lemma 3: If $\widehat{\Gamma}^*$ has a value then $\widehat{\Gamma}$ has a value and $\text{val}(\widehat{\Gamma}^*) = \text{val}(\widehat{\Gamma})$.

Proof: The proof uses the previous lemma. Let

$$\underline{v} = \sup_{\hat{\sigma}} \inf_{\hat{\tau}} H(\hat{\sigma}, \hat{\tau}) \qquad \overline{v} = \inf_{\hat{\tau}} \sup_{\hat{\sigma}} H(\hat{\sigma}, \hat{\tau})$$

where $H(\hat{\sigma}, \hat{\tau})$ is the payoff in the $\widehat{\Gamma}$ game. In a similar way we define $\underline{v}^*$ and $\overline{v}^*$ for the $\widehat{\Gamma}^*$ game. Then, the previous lemma states that:

$$\inf_{\hat{\tau}^*} H(\hat{\sigma}^*, \hat{\tau}^*) = \inf_{\hat{\tau}} H(\hat{\sigma}^*, \hat{\tau})$$

$$\sup_{\hat{\sigma}^*} H(\hat{\sigma}^*, \hat{\tau}^*) = \sup_{\hat{\sigma}} H(\hat{\sigma}, \hat{\tau}^*)$$

Hence,

$$\inf_{\hat{\tau}^*} H(\hat{\sigma}^*, \hat{\tau}^*) = \inf_{\hat{\tau}} H(\hat{\sigma}^*, \hat{\tau}) \leq \underline{v}$$

$$\sup_{\hat{\sigma}^*} H(\hat{\sigma}^*, \hat{\tau}^*) = \sup_{\hat{\sigma}} H(\hat{\sigma}, \hat{\tau}^*) \geq \overline{v}.$$

These imply that: $\underline{v} \geq \underline{v}^*$ and $\overline{v} \leq \overline{v}^*$ and the conclusion follows. ◇

Here we introduce the following *stochastic game*, which we denote by $\widehat{\Gamma}^{**}$.

(i) The action space of I is $A = (P^m)^k$. An "action" (pure strategy) is given by $\sigma = \{s_\ell(i); \ell \in \{1, \ldots, k\}, i \in \{1, \ldots, m\}\}$.

(ii) The action space of II is $B = \{1, \ldots, n\}$.

(iii) The state space is $S = P^k$, the probability simplex in $\mathbf{R}^k$.

(iv) The law of motion is $Q(\cdot \mid p, \sigma, j)$, which is a probability distribution over $S \cup \{\emptyset\}$ giving weight to at most $m+1$ points. The i-th point, $i = 1, \ldots, m$ is given by

$$(p(i))_r = \frac{\sum_{\ell \in S} q(r \mid \ell, i, j) s_\ell(i) p_\ell}{\sum_{r \in S} \sum_{\ell \in S} q(r \mid \ell, i, j) s_\ell(i) p_\ell}$$

and carries weight $\sum_{r \in S} \sum_{\ell \in S} q(r \mid \ell, i, j) s_\ell(i) p_\ell$. The $m+1$-th point is the $\emptyset$ always (end of the game).

(v) The reward (immediate payoff) is defined by:

$$R(p, \sigma, j) = \sum_{i=1}^{m} \sum_{\ell \in S} s_\ell(i) p_\ell G^\ell_{ij}.$$

Remark: Clearly $\widehat{\Gamma}^{**}$ is defined through Γ, the SGLIOS. We will say that $\widehat{\Gamma}^{**}$ is the stochastic game generated from Γ. Notice also that, for all p, σ, j, the stop probabilities $Q(\emptyset \mid p, \sigma, j)$ satisfy

$$Q(\emptyset \mid p, \sigma, j) = 1 - \sum_{i=1}^{m} \sum_{r \in S} \sum_{\ell \in S} q(r \mid \ell, i, j) s_\ell(i) p_\ell > 1 - \beta > 0,$$

since $\sum_{r \in S} q(r \mid \ell, i, j) \leq \beta$ and $\beta < 1$.

Lemma 4: The game $\widehat{\Gamma}^{**}$ is identical to the game $\widehat{\Gamma}^{*}$.

Proof: Obviously, the strategy spaces are the same in both these games. It is also easy to see that the stop probabilities at each step are the same. Hence, it is sufficient to show that the immediate payoff is also the same, i.e.,

$$H_N^{\widehat{\Gamma}^{**}}(\hat{\sigma},\hat{\tau}) = H_N^{\widehat{\Gamma}^{*}}(\hat{\sigma},\hat{\tau}).$$

Equations (8),(9) and (10) lead to:

$$H_N^{\widehat{\Gamma}^{*}}(\hat{\sigma},\hat{\tau}) = \sum_{j^{N-1},\sigma^{N-1},p^N} \Big\{ P(j^{N-1},\sigma^{N-1},p^N \mid T>N) \cdot \sum_{\theta_N,i_N,j_N} \Big[P(i_N \mid \theta_N, j^{N-1},\sigma^{N-1},p^N,T>N) \cdot p_N(\theta_N) P(j_N \mid j^{N-1},\sigma^{N-1},p^N,T>N) G^{\theta_N}_{i_N j_N} \Big] \Big\} \quad (16)$$

From equation (9) we had

$$(E_{\hat{\sigma}\hat{\tau}}[\sigma_N])_{\theta_N}(i_N) = P(i_N \mid \theta_N, j^{N-1},\sigma^{N-1},p^N).$$

Hence equation (16) becomes:

$$H_N^{\widehat{\Gamma}^{*}}(\hat{\sigma},\hat{\tau}) = \sum_{j^{N-1},\sigma^{N-1},p^N} \Big\{ P(j^{N-1},\sigma^{N-1},p^N \mid T>N) \cdot \sum_{j_N} \Big[\hat{\tau}_N(j_N \mid j^{N-1},\sigma^{N-1},p^N,T>N) \cdot \sum_{\theta_N}\sum_{i_N} (E_{\hat{\sigma}\hat{\tau}}[\sigma_N \mid j^{N-1},\sigma^{N-1},p^N,T>N])_{\theta_N}(i_N) p_N(\theta_N) G^{\theta_N}_{i_N j_N} \Big] \Big\}.$$

This leads to,

$$H_N^{\widehat{\Gamma}^{*}}(\hat{\sigma},\hat{\tau}) = E_{\hat{\sigma}\hat{\tau}}[R(p_N,\sigma_N,j_N)]$$

which implies the result. ◇

Theorem 1: If Γ, the SGLIOS with positive stop probabilities, has a value and if $\widehat{\Gamma}^{*}$, the stochastic game generated from Γ, has a value, then, $\mathrm{val}(\Gamma) = \mathrm{val}(\widehat{\Gamma}^{*})$.

Proof: Immediate with the use of Lemma (1) and Lemma (3). ◇

4. The Existence of a Stationary Optimal Strategy for the Informed Player:

We next show that both Γ and Γ^* possess a value and hence we may use the generated stochastic game to establish "stationarity" for the strategies of the informed player in the SGLIOS Γ. Indeed, since the states in $\widehat{\Gamma}^*$ are the updated probabilities p, a strategy $\hat{\sigma}$ (or $\hat{\tau}$) will be stationary in $\widehat{\Gamma}^*$ if it depends only on p. Stationarity in optimal strategies (or ϵ-optimal) in $\widehat{\Gamma}^*$ implies that I has an optimal (or ϵ-optimal) strategy in the SGLIOS Γ which depends only on p - the updated probability which he independently knows. Note that this is not the case for II, since he needs to know σ_N at each stage N of Γ in order to compute p_{N+1}.

In other classes of SGLIOS, it often turns out that the informed player may restrict his attention to such myopic strategies also (e.g. the One Stage Look Ahead Strategies of [5] or [9]).

Obtaining a characterization of SGLIOS with this property would probably be of interest.

Theorem 2: The common value $v(p)$ of Γ and $\widehat{\Gamma}^*$ exists and is a continuous function on the simplex P^k. Moreover, there exists an optimal stationary (myopic) strategy σ_0 for the informed player in Γ, the SGLIOS with positive stop probabilities. There also exists an optimal stationary strategy τ_0 for II in the generated stochastic game $\widehat{\Gamma}^*$ such that,

$$v(p) = \sum_{j=1}^{n} [\tau_0(p)]_j \left\{ R(p, \sigma_0(p), j) + \sum v(\cdot) Q(\cdot \mid p, \sigma_0(p), j) \right\}.$$

Remark: In the context of $\widehat{\Gamma}^*$, σ_0 is a *pure* optimal stationary strategy.

Proof: Let Γ_N be the truncation of the game Γ after N stages and let v_N be its value. Let also $\sigma(N)$ be the strategy of I in Γ which is optimal in Γ_N and arbitrary from stage $N+1$ and on. Then, equation (1) implies that

$$v_N - \frac{\beta^N M}{1-\beta} \leq H(\sigma(N), \tau), \text{and hence,}$$

$$v_N - \frac{\beta^N M}{1-\beta} \leq \sup_{\sigma} \inf_{\tau} H(\sigma, \tau) = \underline{v}. \tag{17}$$

Using a similar argument for II we get

$$v_N + \frac{\beta^N M}{1-\beta} \geq \inf_{\tau} \sup_{\sigma} H(\sigma, \tau) = \overline{v}. \tag{18}$$

Hence, using (17) and (18) for N large enough, we get the existence of the value for Γ. The continuity of v may be obtained using a similar argument and the continuity of v_N (it is also a corollary of the next proof). The rest of the proof depends on a modification of a result of A. Maitra and T. Parthasarathy. They show that for a discounted stochastic game with compact state and action spaces and continuous reward function the value exists, it is continuous and the two players possess stationary optimal strategies, provided that the transition probabilities $Q(\cdot \mid s, a, b)$ satisfy the following property: Whenever $(s_n, a_n, b_n) \overset{n\to\infty}{\longrightarrow} (s, a, b)$ then $Q(\cdot \mid s_n, a_n, b_n) \underset{\text{weakly}}{\overset{n\to\infty}{\longrightarrow}} Q(\cdot \mid s, a, b)$, where s, a, b represent the state and the actions of the two players (see [10]). In their proof, the discount factor is used to establish the usual contraction properties of certain operators, which still hold if we assume positive stop probabilities bounded away from 0 , i.e., $inf_{s,a,b}\, Q(\emptyset \mid s, a, b) > 0$. This is certainly the case with $\hat{\Gamma}^*$, for which it is also easy to check that S, A, and B are compact metric spaces, R is continuous on $S \times A \times B$ and also if $(p_n, \sigma_n, j_n) \overset{n\to\infty}{\longrightarrow} (p_o, \sigma_o, j_o)$ then $Q(\cdot \mid p_n, \sigma_n, j_n) \underset{\text{weakly}}{\overset{n\to\infty}{\longrightarrow}} Q(\cdot \mid p_o, \sigma_o, j_o)$. This establishes the rest of the theorem up to the existence of a *mixed* stationary optimal strategy for I in $\hat{\Gamma}^*$. But, since A is convex and the reward is linear in σ, the optimal stationary strategy of I may be replaced with its expectation, which is a strategy in Γ that depends only on p. $\diamond$

Acknowledgement

Discussions with Professor T. S. Ferguson were beneficial in clarifying the recursive structure of SGLIOS. I thank him.

References

[1] *Shapley, L.S.* (1953). Stochastic Games, *Proceedings of the National Academy of Sciences*, **39**, No. 11, 1095-1100.

[2] *Feldman, D.* (1962). Contributions to the Two Armed Bandit Problem, *Annals of Mathematical Statistics*, **33**, 847-856.

[3] *Dynkin, E. and A. Yushkevich* (1979). " Controlled Markov Processes", Springer-Verlag.

[4] *Aumann, R. and M. Maschler* (1966). Game Theoretic Aspects of Gradual Disarmament, Report Under Contract ACDA/ST-80, Chapter V.

[5] *Melolidakis, C.* (1984). On Stochastic Games with Lack of Information on One Side, Ph.D. Thesis, UCLA.

[6] *Melolidakis, C.* (1987). On the Existence of Myopic Optimal Strategies for the Informed Player in Stochastic Games with Lack of Information on One Side, Technical Report, TUC.

[7] *Aumann, R.* (1964). Mixed And Behavior Strategies In Infinite Extensive Games, *Annals Of Mathematical Studies*, **52**, 627-650.

[8] *Melolidakis, C.* (1986). On the Recursive Structure of Stochastic Games with Lack of Information on One Side, Technical Report, University of Crete.

[9] *Melolidakis, C.*, On Stochastic Games With Lack of Information on One Side, to appear in *International Journal of Game Theory.*

[10] *Maitra, A. and T. Parthasarathy* (1970). On Stochastic Games, *Journal Of Optimization Theory And Applications*, **5**, 289-300.

[11] *Mertens, J.F.* (1986). Repeated Games, *C.O.R.E. Discussion Paper.*

[12] *Sorin, S.* (1984). Big Match with Lack of Information on One Side, *International Journal Of Game Theory*, **13**, 201-255.

[13] *Sorin, S.* (1985). "Big Match", with Lack of Information on One Side (Part II), *International Journal Of Game Theory*, **14**, 173-204.

Costis Melolidakis
Department of Industrial Engineering
Technical University of Crete
73100 Khania, Crete
Greece

Positive Stochastic Games and a Theorem of Ornstein

Andrzej S.Nowak and T.E.S. Raghavan*

Stochastic games were first formulated by Shapley in 1953. In his fundamental paper Shapley [13] established the existence of value and optimal stationary strategies for zero-sum β-discounted stochastic games with finitely many states and actions for the two players.

A positive stochastic game with countable state space and finite action spaces consists of the following objects:

1. State space S- the set of nonnegative integers.
2. Finite action spaces $A(B)$ for players I(II).
3. The space of mixed strategies $P(A)(P(B))$ on the action spaces $A(B)$ for players I(II).
4. Nonnegative (immediate) reward function $r(s, a, b)$.
5. Markovian transition $q(t|s, a, b)$ where $q(t|s, a, b)$ is the chance of moving from state s to state t when actions a, b are chosen by players I and II in the current state s.

When playing a stochastic game, the players, in selecting their actions for the kth day, could use all the available information to them till that day, namely on the partial history up to the kth day given by $(s_1, a_1, b_1, s_2, a_2, b_2, \ldots s_{k-1}, a_{k-1}, b_{k-1}, s_k)$. Thus a strategy φ for player I is a sequence $(\varphi_1, \varphi_2, ..)$ where φ_k selects a mixed strategy (an element of $P(A)$), for the kth day.

We can classify stochastic games with additional structure on the immediate rewards and transition probabilities. The law of motion is said to be *controlled by one player* (say player II) if $q(t|s, i, j) = q(t|s, j)$ for all i. We call a stochastic

*This research was partially supported by the N.S.F. Grant DMS 8601403

game, a *switching control* stochastic game if the law of motion $q(t|s,i,j)$ satisfies for some set $S_1 \subset S$

$$q(t|s,i,j) = q(t|s,i) \quad \text{for all } j \text{ if } s \in S_1.$$
$$q(t|s,i,j) = q(t|s,j) \quad \text{for all } i \text{ if } s \notin S_1.$$

We call a stochastic game *positive* if the immediate rewards $r(s,a,b)$ are nonnegative for all s,a,b.

We will in general be interested in two simpler classes of strategies. A strategy φ is called a *stationary strategy for player I* if $\varphi = (f, f, ...)$ where $f : S \to P(A)$. Stationary strategies are similarly defined for player II. Intuitively, a stationary strategy depends only on the current state and not on the route through which the current state is reached. Suppose the game starts initially at state t. A strategy φ for player I is called *semi-stationary* if $\varphi = (f^t, f^t, ...)$ where $f^t : S \to P(A)$. Unlike stationary strategies a semi-stationary strategy depends not only on the current state but also on the initial state t in selecting a randomized action.

In a positive stochastic game, for each starting state t, the expected payoff using strategies φ and γ by players I and II respectively is given by

$$E(\varphi,\gamma)(t) = \sum_{n=1}^{\infty} r_n(\varphi,\gamma)(t)$$

where $r_n(\varphi,\gamma)(t)$ is the expected income on the nth day for player I when the game starts at state t. In this paper we will assume that

$$\sup_{\varphi,\gamma} E(\varphi,\gamma)(s) \le c < \infty \text{ for each state } s. \tag{A}$$

(Here the supremum ranges over arbitrary strategies φ and γ for the two players).

For any stationary strategy $\varphi = (f, f, ...)$ for player I and for any stationary strategy $\gamma = (g, g, ...)$ for player II , we can associate the well known dynamic programming operator L_{fg} where for any nonnegative function u on S

$$(L_{fg}u)(s) = \sum_{a\in A}\sum_{b\in B}[r(s,a,b) + \sum_{t\in S} u(t)q(t|s,a,b)]f(a|s)g(b|s), s \in S.$$

From Blackwell [1], we know that (A) holds if and only if for some bounded nonnegative function u and for all actions $a \in A$ and $b \in B$

$$(L_{ab}u)(s) = r(s,a,b) + \sum_{t\in S} q(t|s,a,b)u(t) \le u(s) \text{ for all } s \in S.$$

The following result was established (in various settings) in [5], [6],[7], [8],[9] and [11].

Theorem 1: Every positive stochastic game has a value and player II has an optimal stationary strategy $\gamma^* = (g^*, g^*, ..)$. Further the value function v is the smallest nonnegative solution to the equation

$$v(s) = \min_{\mu \in P(B)} \max_{\lambda \in P(A)} (L_{\lambda\mu} v)(s), s \in S. \tag{1}$$

Moreover,

$$v(s) = \max_{\lambda \in P(A)} (L_{\lambda g^*} v)(s), \text{ for all } s \in S. \tag{2}$$

(Here $L_{\lambda g^*}$ is defined in the obvious fashion. Furthermore for any $\epsilon > 0$, player I has a semi-stationary ϵ-optimal strategy $\varphi^t = (f^t, f^t, ...)$ for each starting state t.

Blackwell [1] showed in the one person games (positive dynamic programming) that when the value function is unbounded there may not be any ϵ-optimal pure stationary strategy for player I. Van der Wal modified his example and showed [14] that there may not exist any ϵ-optimal mixed stationary strategies, for the positive dynamic programming problems with unbounded value. Kumar and Shiau [7] showed that there may not be optimal strategies (stationary or otherwise) for player I even when the state and action spaces are finite.

In an elegant paper [10], Ornstein showed that if the optimal return function v is bounded and if the state space is countable then in positive dynamic programming problems the optimizer has an ϵ-optimal pure stationary strategy. Van Dawen [4] pointed out that if player I controls the law of motion then Ornstein's theorem can be extended to positive stochastic games. Different sufficient conditions were given in [9] (Theorem 5.5) and [14] for the existence of ϵ-optimal stationary strategies.

Our main result of this paper is a counterexample which shows that Ornstein's theorem cannot be extended to positive stochastic games with general transition functions.

Counter-example: Let S = 0, 1, 2,... be the states of the system. Let state 0 be absorbing. Let players I and II have two actions in each state.

Case (i). At states $s \geq 2$ player I chooses action 1.

As long as $s \geq 2$ there is no immediate reward to player I. However the game moves to state $s - 1$ if player II chooses action 1 and moves to state $s + 1$ if player II chooses action 2.

Case (ii). At state 1 player I receives 1 from player II, no matter what actions are chosen by the players and the game moves to the absorbing state 0 where, player I receives nothing from then on.

Case (iii). At state $s \geq 2$ player I chooses action 2. Starting at state s if player II chooses action 1, then the immediate reward is 0 and the game moves to the absorbing state 0. If player II chooses action 2 then the immediate reward is 1 and the game moves to the absorbing state 0.

We will show that for this game there is no ϵ-optimal stationary strategy for player I. We need the following lemma in the sequel whose proof, though classical and known for special cases is given for the sake of completeness.

Lemma 1: Consider a random walk on the set of nonnegative integers where state t is absorbing. At state s let the process move one step to the right with chance $\xi(s)$ and move one step to the left with chance $\eta(s)$ and $\xi(s)+\eta(s)=1$. Let $\xi(s)\leq\frac{1}{2}$ for all states $s>t$. If the process starts at a state $s>t$ then the process will be absorbed at state t with probability 1.

Proof: For any starting state $s>t$ of the random walk, let $N>s$ be a state to the right of s. Consider a modified random walk which has another absorbing state $N>s>t$. For the modified random walk let π_s^N be the probability of absorption to state N starting at s. Clearly from the symmetric random walk of the Gambler's Ruin problem [12] we have $1-\pi_s^N\to 1$ as $N\to\infty$. Our original random walk starting at a state $s>t$ has at least as much chance as the modified random walk to get absorbed at t for all $N>s$ and we have the absorption probability $\geq 1-\pi_s^N$ for all N. Thus the random walk gets absorbed at state t with probability 1. ◇

It is easily seen that $E(\varphi,\gamma)(s)\leq 1$ for all strategies φ,γ and for all states s. By Theorem 1 the value of the positive stochastic game is the smallest nonnegative solution to the equation

$$v(s)=\text{value}\begin{bmatrix} v(s-1) & v(s+1) \\ 0 & 1 \end{bmatrix}, s\geq 2. \tag{3}$$

Further $v(0)=0$ and $v(1)=1$.

Since by choosing the first action in each state s, player II loses at most $v(s-1)$ the value of the game satisfies the monotonicity condition $v(s)\leq v(s-1)$ for all states $s\geq 2$.

Lemma 2: $\lim_s v(s)=\nu$ exists and $\geq\frac{1}{2}$.

Proof: Since $v(s)$ is nonnegative and nonincreasing $\lim_s v(s)$ exists. We have $v(1)=1, v(3)\geq 0$ and

$$v(2)=\text{ value }\begin{bmatrix} 1 & v(3) \\ 0 & 1 \end{bmatrix}\geq\frac{1}{2}. \tag{4}$$

Suppose, $\nu<\frac{1}{2}$. By (4) we therefore have a state $t\geq 2$ with the largest index such that $v(t)\geq\frac{1}{2}$. Assume that the game starts at some state $s>t$. Let π be any strategy for player I which selects action 1 in all states $s>t$ and switches over to an ϵ-optimal semi-stationary strategy π^t, once state t is reached. By Theorem 1, and from (3), for any optimal stationary strategy $\gamma^* = (g^*, g^*, ..)$ of player II with $g^*(s) = (g_1^*(s), g_2^*(s))$, we have

$$0.g_1^*(s) + 1.g_2^*(s) \leq v(s) < \frac{1}{2}$$

Thus $g_2^*(s) < \frac{1}{2}$ for all the states $s > t$ and therefore $g_1^*(s) > \frac{1}{2}$ for all $s > t$. When player I uses the above mentioned strategy π and player II uses the optimal stationary strategy γ^*, by the above Random Walk Lemma the game will reach state t with probability 1 starting at any state $s > t$. We have

$$E(\pi, \gamma^*)(s) = E(\pi^t, \gamma^*)(t) \geq v(t) - \epsilon$$

and $v(s) = sup_\varphi E(\varphi, \gamma^*)(s) \geq v(t) - \epsilon$. Since $\epsilon > 0$ is arbitrary we have $v(s) \geq v(t)$ for all $s > t$. This contradicts our assumption about state t. ◇

We will in fact give the exact value of $v(s)$ in the following lemma.

Lemma 3: The value $v(s) = \frac{(s+1)}{2s}$ for all s.

Proof: We know from Theorem 1 that the value function is the smallest nonnegative solution to (3). Clearly $v(s) = \frac{(s+1)}{2s}$ satisfies equation (3). Suppose it is not the smallest solution to equation (3). Let k be the first index for which $v(k)$ is $< \frac{(k+1)}{2k}$. Clearly $k > 1$. Let $v(k) = \frac{(k+1)}{2k} - \epsilon$ where $\epsilon > 0$. We will prove that $v(k+1) = \frac{(k+2)}{2(k+1)} - \delta$ where $\delta \geq \epsilon$. From (3) we have

$$v(k) = \text{ value } \begin{bmatrix} v(k-1) & v(k+1) \\ 0 & 1 \end{bmatrix}$$

Since the game matrix above is completely mixed the formula for the value can be used and we get

$$\frac{(k+1)}{2k} - \epsilon = \frac{\frac{k}{2(k-1)}}{1 + \frac{k}{2(k-1)} - v(k+1)}.$$

Now solving for $v(k+1)$ we get

$$v(k+1) = \frac{3k-2}{2(k-1)} - \frac{k^2}{(k^2-1)} \cdot \frac{1}{(1 - \frac{2\epsilon k}{k+1})}.$$

(Here we use the validity of the formula for $v(k-1)$ in computing the value $v(k+1)$.) Since $0 < v(k)$ we have $\frac{2\epsilon k}{(k+1)} < 1$ and expanding the last term in a power series we get

$$\begin{aligned} v(k+1) &= \frac{3k-2}{2(k-1)} - \frac{k^2}{(k^2-1)}(1 + \frac{2\epsilon k}{(k+1)} + \text{positive terms}) \\ &= \frac{k+2}{2(k+1)} - \frac{2\epsilon k^3}{(k^2-1)(k+1)} - 0(\epsilon^2) - ... \end{aligned}$$

Since $k > 1$ we have $\frac{k^3}{(k^2-1)(k+1)} > \frac{1}{2}$. Thus $v(k+1) = \frac{k+2}{2(k+1)} - \delta$ where $\delta \geq \epsilon$. Continuing this way we will find for all $N, v(N) = \frac{(N+1)}{2N} - \delta_N$ where $\delta_N \geq \epsilon$. This contradicts the previous lemma that the limiting value $\nu \geq \frac{1}{2}$. ◇

Lemma 4: There exists no stationary ϵ-optimal strategy for player I.

Proof: Let $\varphi = (f, f, ...)$ be any stationary strategy for player I. Let $f(s) = (f_1(s), f_2(s))$ where $f_i(s)$ is the chance of selecting action $i = 1, 2$ at state s.

Case 1: $\sum_{s \geq 0} f_2(s) < \infty$.

Let γ_2 be the strategy where player II uses action 2 at all states.

We have

$$\begin{aligned} E(\varphi, \gamma_2)(s) &= f_2(s) + f_1(s) f_2(s+1) + f_1(s) f_1(s+1) f_2(s+2) + ... \\ &\leq \sum_{k=s}^{\infty} f_2(k) \to 0 \text{ as } s \to \infty. \end{aligned}$$

Thus for small $\epsilon > 0$ and large s such a φ cannot guarantee player I an expected value $\geq v(s) - \epsilon$ and therefore φ cannot be ϵ-optimal.

Case 2: $\sum_{s \geq 0} f_2(s) = \infty$.

Let γ_1 be the stationary strategy where player II chooses action 1 at all states. We have

$E(\varphi, \gamma_1)(s) =$ probability of reaching state 1 $= f_1(s).f_1(s-1)...f_1(2)$. Thus $\prod_{k=2}^{s} f_1(k) = \prod_{k=2}^{s}(1 - f_2(k)) \leq \frac{1}{1+\sum_2^s f_2(k)} \to 0$ as s $\to \infty$ (By the Weierstrass Inequality [3].)

We again have the assertion that φ cannot guarantee player I an expected value $\geq v(s) - \epsilon$ for large states s. Thus there are no stationary ϵ-optimals for player I. This completes the proof of the lemma and the proof of the required assertion. ◇

Remark: It is therefore natural to look for subclasses of positive stochastic games that admit ϵ-optimal stationary strategies for player I. Besides one player control stochastic games, some other natural subclasses are the switching control stochastic games and irreducible stochastic games. For the case of player I control stochastic games the theorem of Ornstein extends in an obvious way. For Player II control positive stochastic games one can prove the stronger result that player I has an optimal stationary strategy. Besides one player control games one can also extend Ornstein's proof to prove the the following

Theorem: Let G be a switching control positive stochastic game with a countable state space S and finite action spaces. Let G satisfy condition (A). Let S_1 be the states controlled by player I. For all actions j of player II let $q(S_1|s,j) = 0$ for all states $s \notin S_1$. Then the positive stochastic game admits an ϵ-optimal stationary strategy for player I.

A formal proof of this theorem will be published elsewhere.

Acknowledgement: We wish to thank Professor T.S. Ferguson for some stimulating discussions on this problem. He basically gave us the formula for the value in Lemma 3 and also suggested a possible proof. Needless to say, that our faith in the existence of a counter example was inspired by the game of Big Match [2].

References

[1] *Blackwell, D.* (1967). Positive Dynamic Programming. *Proc. Fifth Berkeley Symp. Math. Stat. Prob.*, **2**, 415-418. University of California Press.

[2] *Blackwell, D. and T.S. Ferguson* (1968). The Big Match, *Ann. Math. Stat.*, **39**, 159-163.

[3] *Bromwich, T.J.I.A.* (1926). "An Introduction to the Theory of Infinite Series", 2nd Edition, McMillan and Company, London.

[4] *Van Dawen, R.* (1984). On Stationary Strategies in Positive Stochastic One and Two Person Games with General State Space. *Z. Angew. Math. Mech.*, **64**, 327-328.

[5] *Frid, E. B.* (1973). On Stochastic Games. *Theory Probability and Appl.*, **18**, 389-393.

[6] *Himmelberg, C.J., T. Parthasarathy, T.E.S. Raghavan and F.S. Van Vleck* (1976). Existence of p-Equilibrium and Optimal Stationary Strategies in Stochastic Games. *Proc. Amer. Math. Soc.*, **60**, 245-251.

[7] *Kamerud. D.B.* (1975). Repeated Games and Positive Stochastic Games, Ph.D. Dissertation, University of Minnesota.

[8] *Kumar, P.R. and T.H. Shiau* (1981). Existence of Value and Randomized Strategies in Zero-Sum Discrete Time Stochastic Dynamic games. *SIAM J. Control Optim.*, **19**, 617-634.

[9] *Nowak, A.S.* (1985). Universally Measurable Strategies in Zero-Sum Stochastic Games. *Ann. Prob.*, **13**, 269-287.

[10] *Ornstein, D.* (1969). On the Existence of Stationary Optimal Strategies. *Proc. Amer. Math. Soc.*, **20**, 563-569.

[11] *Rieder, U.* (1978). On Semi-Continuous Dynamic Games. Technical Report, University of Karlsruhe.

[12] *Ross, S.M.* (1981). Introduction to Probability Models. Academic Press, New York.

[13] *Shapley, L.S.* (1953). Stochastic Games, *Proc. Nat. Acad. Sci., U.S.A.*, **39**, 1095-1100.

[14] *Van der Wal, J.* (1981). Stochastic Dynamic Programming. *Math. Centre Tract*, **139**, Mathematische Centrum, Amsterdam.

A.S.Nowak
Instytut Mathematyki
Polytechnika Wroclawska
Wybrzeze Wyspianskiego 27
50-370 Wroclaw
Poland

T.E.S. Raghavan
Department of Mathematics, Statistics & Computer Science
University of Illinois at Chicago
Chicago, Illinois 60680-4348
U.S.A.

Existence of Correlated Weak Equilibria in Discounted Stochastic Games with General State Space*

Andrzej S. Nowak

Abstract: This paper treats of nonzero-sum discounted stochastic games with general state space where the players are allowed to use correlated strategies. The concept of correlated weak equilibrium of Moulin and Vial (being an extension of the Nash equilibrium point) is adopted to such games. An existence theorem is proved for a class of discounted stochastic games with a Borel state space where stationary Nash equilibria are not known to exist.

1. Introduction: The concept of weak equilibrium in randomized correlated strategies for nonzero-sum games was introduced by Moulin and Vial [9] as "simple extension" of Aumann's correlated equilibrium [1]. It is apparent from the definitions that the set of all correlated weak equilibria in a matrix game is a compact convex set containing the set of correlated equilibria "a la Aumann" and the latter set contains the convex hull of Nash equilibria. For a detailed discussion of these topics including some interesting examples the reader is referred to [8], [9, §1.2] and [1].

In this paper we adopt the correlation scheme of Moulin and Vial [9] to discounted stochastic dynamic games with general state space and compact metric action spaces. The concept of correlated weak equilibrium introduced here for such games is a natural extension of the Nash equilibrium point in stationary strategies and seems to be an interesting solution concept also for finite state nonzero-sum stochastic games. The second objective of this paper is to give some sufficient conditions for the existence of a stationary correlated equilibrium in a general state discounted stochastic game (see (A_1), (A_2) in §2). These conditions are rather mild and our

*This paper was presented at the International Workshop on Stochastic Games organized by Prof. T.E.S. Raghavan at the University of Illinois at Chicago in June 26-28, 1987.

main result is a new existence theorem, because stationary Nash equilibria are known to exist only in some very special cases of our framework. Namely, they have been shown to exist in games with finite action spaces where payoffs and transitions satisfy certain separability conditions [6,12] or transitions are state independent [13]. For more general models only some ϵ-Nash equilibrium theorems are known [10, 14, 17].

2. Main Result: We begin with some preliminaries. Let X be a Borel space, i.e, X is a nonempty Borel subset of a complete separable metric space. We assume that X is given the relative topology and the Borel σ-algebra denoted by $\mathcal{B}(X)$. By $P(X)$ we denote the space of all probability measures on $\mathcal{B}(X)$, endowed with the usual weak topology. Let X and Y be Borel spaces. A transition probability from X to Y is a function $t : \mathcal{B}(Y) \times X \to [0,1]$ such that, for each $B \in \mathcal{B}(Y)$, $t(B|\cdot)$ is a Borel measurable function and, for each $x \in X$, $t(\cdot|x) \in P(Y)$. We denote by $Q(Y|X)$ the set of all transition probabilities from X to Y. It is known that every $t \in Q(Y|X)$ may be recognized as a Borel measurable mapping from X to $P(Y)$, and vice versa [3, Proposition 7.26].

In this paper we consider the standard nonzero-sum discounted stochastic dynamic game model studied in [6, 10-14, 17], but we allow the players to use correlated strategies. Let $N = \{1, \ldots, n\}$ be the *set of players*. An *n-person discounted noncooperative stochastic game* is defined by a sequence of objects: $\{S, A_1, \ldots, A_n, q, r_1, \ldots, r_n, \beta\}$ where:

(i) S is a Borel space, the *state space* for the game.

(ii) A_k is a nonempty compact metric space, the *action space* for player $k \in N$.

(iii) q is a transition probability from $S \times A$ to S, the *transition law* for the model. Here $A = \prod_{k \in N} A_k$ is the Cartesian product of A_k's, endowed with the product topology.

(iv) $r_k : S \times A \to R$ is a bounded Borel measurable *payoff function* for player $k \in N$.

(v) $\beta \in (0,1)$ and is called the *discount factor*.

Such a game is played by stages. We assume that at any stage the players know the entire history of the game and are allowed to correlate their choices according to the scheme introduced (for static games) by Moulin and Vial [9, §1.2] as "simple extension" of Aumann's correlation scheme [1] (see also [8]). Every player wants to maximize his expected discounted income. Our main results states that, under some additional assumptions (see (A_1), (A_2) below) our game possesses a stationary correlated weak equilibrium. In view of this, in this paper we will not consider nonstationary and history dependent correlated strategies for the players. A *stationary correlated strategy* for the players is a sequence $f = (\varphi, \varphi, \ldots)$ where $\varphi \in Q(A|S)$. Let F stands for the *set of all stationary correlated strategies* for the players. For any $f \in F$ and any initial state $s \in S$, the *β-discounted expected income for player* $k \in N$ is

$$I_k(f)(s) = \sum_{m=1}^{\infty} \beta^{m-1} r_k^m(f)(s).$$

Here $r_k^m(f)(s)$ is the mth day expected income for $k \in N$ (see [4] or [10, §3] for details).

Let A^{-k} be the Cartesian product $\prod_{m \in N-\{k\}} A_m$, $k \in N$. Then A^{-k} is the set of all actions played by every one but player $k \in N$. Define $F^{-k} = \{(\varphi, \varphi, \ldots) : \varphi \in Q(A^{-k}|S)\}$, $k \in N$. Then F^{-k} is the set of stationary correlated strategies for the players from $N - \{k\}$.

Let π^{-k} be the projection mapping from A onto A^{-k}. For each Borel set $B \in \mathcal{B}(A^{-k})$, we put $B \times A_k = (\pi^{-k})^{-1}(B)$. Then $B \times A_k$ is a Borel subset of A. Let $f \in Q(A|S)$. Associated with f and $k \in N$ is a marginal Borel measurable transition probability f^{-k} from S to A^{-k}. Namely, $f^{-k}(B|s) = f(B \times A_k|s)$, where $B \in \mathcal{B}(A^{-k})$, $s \in S$. Clearly, if $(f, f, \ldots,) \in F$, then $(f^{-k}, f^{-k}, \ldots) \in F^{-k}$.

Let F_k be the *set of all stationary strategies for player* $k \in N$. Of course, every element of F_k is a sequence $g = (\gamma, \gamma, \ldots)$, where $\gamma \in Q(A_k|S)$. For any $g \in F_k$, $\varphi \in F^{-k}$ and $s \in S$, we use $I_k(\varphi, g)(s)$ to denote the discounted expected income for player k, when he uses g while the remaining players use a correlated strategy φ (for a formal definition of $I_k(\varphi, g)(s)$ consult [4] or [10, §3]). Now we can introduce the solution concept that is studied in this paper.

We say that $f \in F$ is a *correlated weak equilibrium* for the game described above if for each player $k \in N$, we have

$$I_k(f)(s) \geq I_k(f^{-k}, g)(s) \quad \text{for all } g \in F_k \text{ and } s \in S. \tag{1}$$

The interpretation is as follows: At first a correlated strategy $f \in F$ is fixed during a preplay communication process between the players. Then f is sent along to a neutral arbiter who realizes the game. It is assumed that after choosing f but before starting the game every player has the right to change his mind (revise his previous decision) and he may ask the arbiter to employ on his behalf some of his private strategies. (No communication between the players is allowed at this moment). The stochastic game starts after everyone has made his final decision. If player k insists on using his strategy $g \in F_k$ while the remaining players decide to correlate their actions during the play according to $f \in F$, fixed at the beginning, then the expected income for player k will be $I_k(f^{-k}, g)(s)$. In view of the above, (1) says that if f is a correlated weak equilibrium, then unilateral deviations from it are not profitable. From the mathematical point of view, (1) is an extension of the Nash equilibrium concept in stationary strategies (see Remark 1).

To ensure the existence of a correlated weak equilibrium for our game, we make some additional assumptions. Let $C(A)$ be the space of all continuous real-valued functions on A, endowed with the supremum norm $\|\cdot\|$. We assume that:

(A_1) $r_k(s,\cdot) \in C(A)$, for each $k \in N$, $s \in S$.

(A_2) q has a Borel measurable density function $b : S \times S \times A \to R$ with respect to some $p \in P(S)$ and b satisfies the following conditions:

(a) $b(s,t,\cdot) \in C(A)$, for each s, $t \in S$, and

(b) $\int \|b(s,t,\cdot)\|p(dt) < \infty$, for each $s \in S$.

Here is the main result of this paper.

Equilibrium Theorem: Any nonzero-sum discounted stochastic game satisfying (i)-(v) and $(A_1), (A_2)$ has a stationary correlated weak equilibrium.

Remark 1: Let $f = (\varphi, \varphi, \ldots)$ be a stationary correlated strategy for the players such that, for each $s \in S$, $\varphi(\cdot|s)$ is a product probability measure on A. If f satisfies (1), then it is a stationary Nash equilibrium for the game. Thus, the set of stationary weak equilibria for a game satisfying (A_1) and (A_2) is broader than the set of stationary Nash equilibria. However, in stationary strategies only ϵ-Nash equilibria ($\epsilon > 0$) are known to exist for games satisfying (A_1) and (A_2) [10]. The question as to whether such games possess stationary Nash equilibria is still open. Some partial solutions to this problem (requiring certain additional assumptions) may be found in [6,12,11,13]. In this paper we do not discuss the rich theory of zero-sum stochastic games started in 1953 by Shapley [15] and developed afterwards by many authors.

Remark 2: Suppose $f \in F$ is a weak equilibrium for a game satisfying (i) - (v). Then unilateral deviations from f using history dependent (alias nonstationary) strategies are not profitable as well. This is a corollary to Blackwell's classical work on discounted dynamic programming [4, Theorem 6].

3. Proof of the equilibrium theorem: Throughout this section p is a fixed probability measure from $P(S)$ dominating the transition probabilities $q(\cdot|s,a)$, $s \in S$, $a \in A$. Let $\overline{Q}(A|S)$ be the space of all classes of p -- a.e. equal transition probabilities from $Q(A|S)$. Similarly, we define $\overline{Q}(A^{-k}|S)$. Let $L_1 = L_1(S,p,C(A))$ denote the Banach space of all $C(A)$-valued p integrable functions defined on S with the norm $\|\cdot\|_1$ given by

$$\|u\|_1 = \int \|u(s,\cdot)\|p(ds), \quad u \in L_1,$$

$\|\cdot\|$ means the supremum norm in $C(A)$. It is well-known that $\overline{Q}(A|S)$ is a closed part of the unit ball of the topological dual space L_1^* endowed with the weak-star topology [7, VII.7]. By the Banach-Alaoglu theorem [5, Theorem V.4.2], $\overline{Q}(A|S)$ is compact in the weak-star topology. Because the Borel σ-algebra in S is countably generated, L_1 is separable and consequently, $\overline{Q}(A|S)$ is a metrizable space. Similarly, we infer that $\overline{Q}(A^{-k}|S)$ is a compact metrizable space.

A Carathéodory integrand on $S \times A$ is a function $u : S \times A \rightarrow R$ such that $u(s,\cdot) \in C(A)$, for each $s \in S$, $u(\cdot, a)$ is Borel measurable for $a \in A$, and $\int \|u(s,\cdot)\| p(ds) < \infty$. It is easy to see that $\varphi_m \longrightarrow \varphi$ in $\overline{Q}(A|S)$, equipped with the weak-star topology, if and only if for every Carathéodory integrand u, we have

$$\int \left[\int u(s,a)\varphi_m(da|s) \right] p(ds) \longrightarrow \int \left[\int u(s,a)\varphi(da|s) \right] p(ds).$$

It is also obvious that if $\varphi_m \longrightarrow \varphi$ in $\overline{Q}(A|S)$, then for each $k \in N$, $\varphi_m^{-k} \longrightarrow \varphi^{-k}$ in $\overline{Q}(A^{-k}|S)$.

Let $B(S)$ be the space of all bounded Borel measurable real-valued functions on S. We identify all functions from $B(S)$ that are p – a.e. equal and the quotient space in such a way obtained we denote by $\overline{B}(S)$. We say that $\{u_m\}$ converges weakly in $\overline{B}(S)$ to some u if, for every $C \in \mathcal{B}(S)$, $E1_C u_m \underset{m}{\longrightarrow} E1_C u$ as $m \rightarrow \infty$. Here and in the sequel E means the expectation operator with respect to p.

We now introduce some auxiliary operators that are well-known in discounted dynamic programming [4] and stochastic games (see e.g., [6, 10, 11, 14]). Let $u \in B(S)$ and $\varphi \in Q(A|S)$. For each $k \in N$, we put

$$(L_\varphi^k u)(s) = \int \left[r_k(s,a) + \beta \int u(t) q(dt|s,a) \right] \varphi(da|s), \quad s \in S,$$

and

$$(L_{\varphi^{-k}\gamma}^k u)(s) = \iint \left[r_k(s,a) + \beta \int u(t) q(dt|s,a) \right] \varphi^{-k}(da^{-k}|s)\gamma(da_k|s)$$

where $s \in S$, $\gamma \in Q(A_k|S)$. If $\gamma(\cdot|s) = \delta_a$ is the Dirac measure at some $a \in A_k$ for all $s \in S$, then we write $(L_{\varphi^{-k}\gamma}^k u)(s)$ as $(L_{\varphi^{-k}a}^k u)(s)$.

The interpretation of $(L_\varphi^k u)(s)$ is as follows: this is the expected income for player k, as a function of the initial state, when the players start using φ, but are terminated at the beginning of the second stage with a final payoff $u(t)$, where t is the state of termination. The operators above are defined for functions and transition probabilities, but they can be (in an obvious way) extended to elements of $\overline{B}(S)$ and $\overline{Q}(A|S)$, $\overline{Q}(A^{-k}|S)$. Taking this into account we can state the following auxiliary result.

Lemma: Let $\{u_m\}$ be a sequence in $\overline{B}(S)$ such that $sup_m |u_m| < \delta < \infty$ p-a.e. Assume that $u_m \longrightarrow u$ weakly in $\overline{B}(S)$ and $\varphi_m \longrightarrow \varphi$ in $\overline{Q}(A|S)$. Then $L^k_{\varphi_m} u_m \longrightarrow L^k_{\varphi} u$ weakly in $\overline{B}(S)$, that is, for every $C \in \mathcal{B}(S)$, we have

$$E1_C L^k_{\varphi_m} u_m \underset{m}{\longrightarrow} E1_C L^k_{\varphi} u \quad \text{as} \quad m \to \infty.$$

Proof: This follows the same lines as Lemma 3.1 in [11]. Similar result holds when φ_m from $\overline{Q}(A|S)$ is replaced by $\varphi_m^{-k} a$, where $\varphi_m^{-k} \longrightarrow \varphi^{-k}$ in $\overline{Q}(A^{-k}|S)$, $a \in A_k$.

Proof of Equilibrium Theorem The starting point is Theorem 3.1 from [10] (see also Remark 5.2 in [10]) stating that our game has a stationary $(1/m)$-Nash equilibrium $\eta_m = (f_1^m, \ldots, f_n^m)$ for each $m \geq 1$. (Clearly, $f_k^m \in F_k$ for each $k \in N$.) Put $v_{km} = I_k(\eta_m)$, the equilibrium payoff for player k corresponding to η_m, and $f_k^m = (\varphi_{km}, \varphi_{km}, \ldots)$, where φ_{km} is some member of $Q(A_k|S)$. Let $\varphi_m[\varphi_m^{-k}]$ be the transition probability from $Q(A|S)[Q(A^{-k}|S)]$ defined as the product of φ_{im}'s, where $i \in N$ $[i \in N - \{k\}]$. For each $s \in S$, $\varphi_m(\cdot|s)[\varphi_m^{-k}(\cdot|s)]$ is the product of measures $\varphi_{im}(\cdot|s)$, $i \in N$ $[i \in N - \{k\}]$.

Note that $v_{km} = L^k_{\varphi_m} v_{km}$ for each player $k \in N$. This follows from the interpretation of $L^k_{\varphi_m} v_{km}$. Suppose now that player k employs a pure action $a \in A_k$ at the beginning of the game and next he plays using $f_{km} \in F_k$. Assuming that the remaining players always use their $(1/m)$-equilibrium strategies the expected payoff to player k is $L^k_{\varphi_m^{-k} a} v_{km}$. Thus, for each initial state $s \in S$, we have

$$1/m + v_{km}(s) = 1/m + (L^k_{\varphi_m} v_{km})(s) \geq (L^k_{\varphi_m^{-k} a} v_{km})(s). \tag{2}$$

We know that $\{v_{km}\}$ is uniformly bounded in $\overline{B}(S)$. (This follows from (iv) and (v).)

By [5, IV.8.4], without loss of generality, $\{v_{km}\}$ has a weak limit in $\overline{B}(S)$, say v_k. At the same time, we can assume that (without loss of generality) $\{\varphi_m\}$ has a limit φ in the weak-star topology on $\overline{Q}(A|S)$. (Recall that $\overline{Q}(A|S)$ is a compact metrizable space.) Then, for each $k \in N$, $\varphi_m^{-k} \longrightarrow \varphi^{-k}$ in $\overline{Q}(A^{-k}|S)$. Using the above lemma and (2), we infer that

$$E1_C v_k = E1_C L^k_{\varphi} v_k \geq E1_C L^k_{\varphi^{-k} a} v_k, \text{ for each } k \in N, \ a \in A_k, \ C \in \mathcal{B}(S). \tag{3}$$

Let D_k be a countable dense subset of A_k. From (3), it follows that for each $k \in N$, there is a Borel set $S_k \subset S$ such that $p(S - S_k) = 0$ and

$$v_k(s) = (L^k_{\varphi} v_k)(s) \geq (L^k_{\varphi^{-k} a} v_k)(s), \quad \text{for each } s \in S_k, \ a \in D_k. \tag{4}$$

Let $T = \bigcap_{k \in N} S_k$. Then $p(S - T) = 0$ and (4) holds for all $s \in T$ and $a \in D_k$, $k \in N$. By our continuity assumptions (A_1) and (A_2), (4) holds for every $s \in T$ and $a \in A_k$, $k \in N$. This implies that, for each $k \in N$ and $\gamma \in Q(A_k|S)$, we have

$$v_k(s) = (L^k_\varphi v_k)(s) \geq (L^k_{\varphi^{-k}\gamma} v_k)(s) \quad \text{for all } s \in T. \tag{5}$$

Now, for each fixed k, consider the discounted dynamic programming model with T as the state space and A_k as the action space. From [4, Theorem 6] and (5), we infer that

$$I_k(f)(s) \geq I_k(f^{-k}, g)(s),$$

where $f = (\varphi, \varphi, \ldots) \in F$, s ranges over T and g is an arbitrary strategy for player k, $g \in F_k$. (Recall that $p(S - T) = 0$ and $q \ll p$). Thus, we have shown that our game has a stationary weak equilibrium for p-a.e. initial state, so-called p-weak equilibrium. In [12], Parthasarathy showed that if a nonzero-sum discounted stochastic game has a p-Nash equilibrium, then it has a Nash equilibrium. Adopting his argument and using Rieder's selection theorem [14], one can obtain a stationary weak equilibrium $f^* = (\varphi^*, \varphi^*, \ldots)$, where φ^* is an appropriate modification of φ on $S - T$. The proof is finished.

Remark 3: One may ask whether the limit φ of $\{\varphi_m\}$ in the above proof determines a Nash equilibrium for our game. In general, that is not the case. The reason is the following: the Cartesian product $\prod_{k \in N} \overline{Q}(A_k|S)$, endowed with the relative topology, is not closed in $\overline{Q}(A|S)$, when p is nonatomic [16, IV.0, IV.2.6]. Moreover, the functionals L^k_φ involved in the proof are not continuous in the product topology on $\prod_{k \in N} \overline{Q}(A_k|S)$. This follows from an obvious modification of Example 2.6 in [2].

Remark 4: A weak equilibrium is obtained in the above proof as a limit of some ϵ-Nash equilibrium points when ϵ tends to zero. Such points do exist by Theorem 3.1 in [10]. It is worth mentioning that one can also get a weak equilibrium for our game using a fixed point argument similar to that one applied in [13] for a different purpose. Weak equilibria in correlated strategies are not studied in [13] explicitly.

Acknowledgement: The author wishes to thank Professor T. Parthasarathy for several useful discussions.

References

[1] *Aumann, R.J.* (1974). Subjectivity and Correlation in Randomized Strategies, *J. Math. Economics* **1**, 67-96.

[2] *Balder, E.J.* (1988). Generalized Equilibrium Results for Games with Incomplete Information, *Math. Oper. Res.* **13**, 265-276.

[3] *Bertsekas, D.P. and S.E. Shreve* (1978). Stochastic Optimal Control: The Discrete Time Case, Academic Press, New York.

[4] *Blackwell, D.* (1965). Discounted Dynamic Programming, *Ann. Math. Statist.* **36**, 226-235.

[5] *Dunford, N. and J.T. Schwartz* (1958). Linear Operators, Vol. I, Interscience Publishers, New York.

[6] *Himmelberg, C.J., T. Parthasarathy, T.E.S. Raghavan and F.S. Van Vleck* (1976). Existence of p-Equilibrium and Optimal Stationary Strategies in Stochastic Games, *Proc. Amer. Math. Soc.* **60**, 245-251.

[7] *Ionescu-Tulcea, A. and C.* (1969). Topics in the Theory of Lifting, Springer-Verlag, Berlin.

[8] *Moulin, H.* (1981). Théorie des Jeux pour L'Économie et la Politique, Hermann, Paris.

[9] *Moulin, H. and J.P. Vial* (1978). Strategically Zero-Sum Games: the Class of Games whose Completely Mixed Equilibria Cannot Be Improved upon, *Internat. J. Game Theory* **7**, 201-221.

[10] *Nowak, A.S.* (1985). Existence of Equilibrium Stationary Strategies in Discounted Noncooperative Stochastic Games with Uncountable State Space, *J. Optim. Theory Appl.* **45**, 591-602.

[11] *Nowak, A.S.* (1987). Nonrandomized Strategy Equilibria in Noncooperative Stochastic Games with Additive Transition and Reward Structure, *J. Optim. Theory Appl.* **52**, 429-441.

[12] *Parthasarathy, T.* (1982). Existence of Equilibrium Stationary Strategies in Discounted Stochastic Games, *Sankhya: The Indian J. Statist. Ser. A* **44**, 114-127.

[13] *Parthasarathy, T. and S. Sinha* (1987). Equilibrium Strategies in Nonzero-Sum Discounted Stochastic Games with Uncountable State Space and State Independent Transitions, Preprint, Indian Statistical Institute, New Delhi.

[14] *Rieder, U.* (1979). Equilibrium Plans for Nonzero-Sum Markov Games, in: Game Theory and Related Topics, Edited by O. Moeschlin and D Pallaschke, North-Holland, Amsterdam.

[15] *Shapley, L.S.* (1953). Stochastic Games, *Proc. Nat. Acad. Sci. U.S.A.* **39**, 1095-1100.

[16] *Warga, J.* (1972). Optimal Control of Differential and Functional Equations, Academic Press, New York.

[17] *Whitt, W.* (1980). Representation and Approximation of Noncooperative Sequential Games, *SIAM J. Control Optim.* **18**, 33-48.

Andrzej S. Nowak
Institute of Mathematics
Wroclaw Technical University
50-370 Wroclaw, Wybrzeze Wyspiańskiego 27
Poland

Nonzero-Sum Stochastic Games[1]

Jean-Francois Mertens and T. Parthasarathy

Abstract: Nonzero-sum discounted stochastic games have equilibrium strategies when the state space is uncountable.

1. Introduction: We will now describe an ℓ-person stochastic game. It is determined by S, A_i, q, r_i, $i = 1, 2, \ldots, \ell$. Here S is a separable measurable space ,the set of states of the system; $A_i = \{1, 2, \ldots, k_i\}$ is a finite set denoting the available actions for player i; $r_i(s, \cdot)$ the immediate rewards to player i which depends on the current state s and the actions chosen by the players, and $q(ds'/s, \cdot)$ is the probability distribution of the next state s'. [Here 'dot' stands for the actions chosen by the ℓ-players]. Then the whole process is repeated from the new state s'. The game is played over the infinite future. Every player wants to maximize his discounted income. The following is the

Main Theorem: Let r_i be bounded and suppose $q(ds'/s, \cdot) \ll \mu$ for every $(s, \cdot)$ where μ is a fixed probability measure. Then there exists a subgame perfect equilibrium where each player's behavioral strategy depends at each date t only on s_t and s_{t-1}.

---

[1] AMS 1980 Subject Classification: Primary 90D15, 28A05, secondary 60K99, 93E05. *Key Words: Stochastic games, Nash equilibrium, selections, dominating measure.*

2. History of the Problem and Known Results: This problem has attracted the attention of several authors, see for example [1,4,5,6,7,8,9]. Every one of these authors gave sufficient conditions, with the exception of [8], for the existence of stationary equilibrium strategies. Himmelberg, Parthasarathy, Raghavan and van Vleck showed that, given any probability distribution p of the initial state, a stationary equilibrium exists for p-almost every initial state, provided the r and q satisfy a strong separability condition and the transition law q is absolutely continuous with respect to p. This result was further strengthened by Parthasarathy [5]. Rieder [7] and Whitt [9] gave sufficient conditions for the existence of ϵ-equilibrium strategies via the approximation of the original game by means of the finite horizon game. Recently Nowak [4] gave sufficient conditions (which are a combination of [5] and [9]) for the existence of ϵ-equilibrium stationary strategies.

In [3], we show that the present result can be strengthened (except for the dependence only on the last two states) by

a) dropping the separability assumption on S.

b) allowing for arbitrary compact action spaces

c) allowing those action sets to vary measurably with the state

d) dropping the requirement that the transition probability be dominated by a measure μ,

and

e) allowing for possibly unbounded reward functions, for discount factors that would be history - and player-dependent, etc.

3. Proof: Let $F_o = \{f : f = (f_1, f_2, \ldots, f_\ell)$ where each $f_i : S \to R$ is measurable and $f_i(s) \in [-c, c]\}$, where c bounds the reward function.

Each entry in F_o may be regarded as possible pay-offs to the players. Given an $f \in F_o$ and an $s \in S$, we define an ℓ-person matrix game $G_f(s)$ where the pay-off to the i-th player is given by $(1-\beta)r_i(s, i, j\ldots) + \beta \int f_i(s')q(ds'|s, i, j\ldots)$. Let $N_f(s) = \{$ Nash equilibrium pay-offs for $G_f(s)\}$. Note $\emptyset \neq N_f(s) \subseteq R^\ell$. Define

$$\overline{N}_f(s) = \text{convex hull of } N_f(s) \text{ if } s \text{ is in non-atomic part of } \mu$$
$$= N_f(s) \text{ if } s \text{ is an atom of } \mu.$$

Let N_f and $\overline{N}_f$ denote all μ-measurable selections from $N_f(s)$ and $\overline{N}_f(s)$ respectively. One can show from any standard selection theorem that such selections do exist. Let $F_1 = \bigcup_{f \in F_o} \overline{N}_f = \overline{N}(F_o)$, and similarly $F_{n+1} = \overline{N}(F_n)$. Observe $F_1 \subseteq F_o$, hence inductively $F_{n+1} \subseteq F_n$.

Lemma: Each F_n is nonempty and w^*-compact (in $L_\infty(\mu)$) for $n = 1, 2, \ldots$

Proof: (by induction). Note F_o is compact and nonempty. Also as observed earlier the correspondence $s \to \overline{N}_f(s)$ has μ-measurable selections. Therefore $F_n \neq \emptyset$ implies $F_{n+1} \neq \emptyset$. In order to complete the proof it is enough to prove that $f_k \xrightarrow{w^*} f$, $\phi_k \in \overline{N}_{f_k}$ and $\phi_k \xrightarrow{w^*} \phi \Longrightarrow \phi \in \overline{N}_f$ i.e., the uppersemicontinuity of $\overline{N}$.

To start with ϕ_k converges on each atom. There exist convex combinations of the ϕ_k's converging to ϕ almost surely. Hence a.s. $\phi(s)$ belongs to the convex hull of the limit points of $\phi_k(s)$. Since $f_k \xrightarrow{w^*} f$, $G_{f_k}(s) \longrightarrow G_f(s)$ for every s. Consequently $\overline{\lim}\, N_{f_k}(s) \subseteq N_f(s)$ everywhere. Hence $\overline{\lim}\, \overline{N}_{f_k}(s) \subseteq \overline{N}_f(s)$. Hence $\phi(s) \in \overline{N}_f(s)$ for every s or $\phi \in \overline{N}_f$. This terminates the proof of the Lemma.

Now define $F = \bigcap_{n=0}^{\infty} F_n$. Clearly $F \neq \emptyset$ as each F_n is compact and $F_n \supseteq F_{n+1}$. From the uppersemicontinuity of $\overline{N}$, it follows that $F = \bigcup_{f \in F} \overline{N}_f$. Choose thus inductively $f_n \in F$ with $f_n \in \overline{N}(f_{n+1})$. Then (Lyapunov) $\int f_n(s_n) q_{s_{n-1}}(ds_n)$ is a measurable map from s_{n-1} to the graph of $\int N_{f_{n+1}}(s_n) q_{s_{n-1}}(ds_n)$. By the measurable choice theorem in [2] it follows thus that there exists a measurable selection $g_n(s_{n-1}, s_n) \in N_{f_{n+1}}(s_n)$, such that $G_{f_n(\cdot)}(s_{n-1}) = G_{g_n(s_{n-1},\cdot)}(s_{n-1})$. [$N_{f_{n+1}}(\cdot)$ is measurable as the composition of the measurable map $s \to G_{f_{n+1}}(s)$ with the equilibrium correspondence, which is by upper semicontinuity a Borel map from games to compact subsets of payoff space]. Denote by $\sigma(p, G)$ a borel selection of an equilibrium with payoff p of the game G. The equilibrium is now obviously to play, at stage n, $\sigma[g_n(s_{n-1}, s_n), G_{f_{n+1}}(s_n)]$.

This proves our theorem.

Comment: If μ is non atomic, the upper semicontinuous correspondence $\overline{N}$ from F_o to itself is convex valued correspondence - hence has a fixed point: $f_o \in \overline{N}_{f_o}$ a.e. Then in the above proof we can choose $f_n = f_o$ for all n, hence $g_n = g$ for all n, hence we get a stationary equilibrium $\sigma(s_{n-1}, s_n)$ instead of $\sigma_n(s_{n-1}, s_n)$).

Acknowledgements: The second author wants to express his sincere thanks to Professor A. Nowak for some useful comments. The second author would also like to thank the Mathematics Institute and CORE at Louvain-La-Neuve for their kind hospitality during his visit from November 7 - December 7, 1985 when this work was initiated.

References

[1] *Himmelberg, C.J., T. Parthasarathy, T.E.S. Raghavan and F. S. Van Vleck* (1976). Existence of p-Equilibrium and Optimal Stationary Strategies in Stochastic Games, *Proc. Amer. Math. Soc.*, **60**, 245-251.

[2] *Mertens, J.F.* (1987). "A Measurable "Measurable Choice" Theorem", CORE Discussion Paper 8749.

[3] *Mertens, J.F., and T. Parthasarathy* (1987). "Equilibria for Discounted Stochastic Games", CORE Discussion Paper 8750.

[4] *Nowak, A.S.* (1985). Existence of Equilibrium Stationary Strategies in Discounted Noncooperative Stochastic Games with Uncountable State Space, *J.O.T.A.*, **54**, 591-602.

[5] *Parthasarathy, T.* (1982). Existence of Equilibrium Stationary Strategies in Discounted Stochastic Games, *Sankhya, Series A*, **44**, 114-127.

[6] *Parthasarathy, T. and S. Sinha* (1987). Existence of Stationary Equilibrium Strategies in Nonzero-Sum Discounted Stochastic Games with Uncountable State Space and State-Independent Transitions, (To appear in *Int. J. Game Theory*).

[7] *Rieder, U.* (1979). Equilibrium Plans for Nonzero-Sum Markov Games, in: O. Moeschlin and D. Pallaschke (eds.) Game Theory and Related Topics, North-Holland, Amsterdam, Holland.

[8] *Sobel, M.J.* (1973). Continuous Stochastic Games, *Jour. Appl. Prob.*, **10**, 497-604.

[9] *Whitt, W.* (1980). Representation and Approximation of Non-Cooperative Sequential Games, *SIAM Jour. on Control and Optimization*, **18**, 33-48.

J. F. Mertens
CORE, 34 Voie du Roman Pays
1348 Louvain-La Neuve
Belgium

T. Parthasarathy
Indian Statistical Institute
7 S.J. Sansanwal Marg
New Delhi-110016
India

On Stochastic Games with Uncountable State and Action Spaces[1]

Rabah Amir[2,3]

ABSTRACT: We consider stochastic games with uncountable state space and prove the existence of a Nash equilibrium in Markovian strategies, reached as a limit, in an appropriate sense, of finite horizon Markovian equilibria (the latter exist even with compact metric action spaces). Our approach is based on Glicksberg's fixed-point theorem, basic measurable selection results and a generalized Fatou Lemma.

1. Introduction: The question of whether every discounted stochastic game with uncountable state space possesses a Nash equilibrium poses a difficult challenge. For finite or countable-state stochastic games, this problem is settled in the affirmative by Shapley [12] for the zero-sum case and Parthasarathy [9] and Federgruen [5] for the nonzero-sum case.

From the standpoint of applications to economic theory in particular, the framework of uncountable state space is often most natural, and an affirmative answer to the question at hand, at least for suitable classes of such games, is highly desirable. Previous attempts have produced partial results. In particular, Himmelberg, Parthasarathy, Raghavan and Van Vleck [7] have an existence theorem, in statio-

[1]*Key Words and Phrases:* stochastic games, stationary and nonstationary dynamic programming, measurable selectors, Glicksberg's fixed-point theorem, generalized Fatou's Lemma.

[2]The research described in this paper was partially supported by ONR Grant No. N00014-86-K-0220.

[3]The author would like to thank E. Balder, J.-F. Mertens and R. Rosenthal for helpful comments, and particularly A. Neyman for many useful discussions.

nary strategies, under restrictive separable additivity assumptions on reward and transition functions. Whitt [15] proves the existence of an ϵ-equilibrium, for every $\epsilon > 0$, by a suitable discretization procedure assuming the basic data of the game to be jointly continuous in states and actions. Nowak [8] strengthens this result by relaxing continuity of the data in the state to Borel measurability. Finally Mertens and Parthasarathy have announced a proof of existence of equilibria in history-dependent strategies.

In this paper, two main results are presented. First, it is shown that every such stochastic game, with uncountable action spaces as well, has a Nash equilibrium in Markovian strategies, provided the horizon is finite (Theorem I). Second, it is shown that if the action spaces are finite and if there exists a finite measure dominating all possible transitions, every such stochastic game has an equilibrium in Markovian strategies when the horizon is infinite (Theorem II).

2. Statement of Problem and Results:

A two-person stochastic game is described by the following objects:

- A state space S, a Borel space (cf. Bertsekas-Shreve [2]).
- Action spaces A_1, A_2 compact metrizable spaces or finite sets (as will be specified).
- One period reward functions r_1, r_2, uniformly bounded non-negative real-valued functions on $S \times (A_1 \times A_2)$, Borel measurable in the first variable and jointly continuous in the other two variables.
- A transition law or stochastic kernel $q(\bullet \mid s, a_1, a_2)$ specifying a probability measure on S for every $(s, a_1, a_2) \in S \times A_1 \times A_2$ such that, for every Borel set $E \in S$, $q(E \mid s, a_1, a_2)$ is Borel measurable in S and jointly continuous on $A_1 \times A_2$.

We consider finite and infinite-horizon stochastic games, both sharing the above description. The horizon is denoted by $T \in N \cup (+\infty)$, where N denotes the natural numbers. A Markovian strategy π is a sequence $(\sigma_t)_{t=0}^{T}$ with σ_t being a Borel map from S into the space $P(A_i)$ of all probability measures on A_i, $t = 0, 1, \ldots, T$. Strategies are either denoted by $\sigma_t(s)$ or $\sigma_t(s; da)$, $s \in S$.

The game unfolds as follows: Upon observing the given initial state $s_1 \in S$, Player i (Pi) selects action $a_1^i \in A_i$ according to a probability measure σ_1^i. The two actions, together with s_1, determine the one-period rewards r_1 and r_2 and the probability distribution $q(\bullet \mid s_1, a_1^1, a_2^1)$ on S according to which the next state s_2 is selected. Then, the entire process is repeated $(T-1)$ more times.

Any pair of Markovian strategies (π_1, π_2) and an initial state $s_1 \in S$ uniquely determine, according to Ionescu Tulcea's theorem, a probability measure

$m(s_1, \pi_1, \pi_2)$ on the space of all possible histories, i.e., on $\{s_1\} \times A_1 \times A_2 \times S_2 \times A_1 \times A_2 \times \times \ldots$. The total discounted expected reward for Pi is defined as

$$I_i(\pi_1, \pi_2) = \int \sum_{t=1}^{T} \beta_i^{t-1} r_i(s_t, a_t^1, a_t^2) dm(s_1, \pi_1, \pi_2), \quad 0 \leq \beta_i < 1.$$

Let $m_t(s_1, \pi_1, \pi_2)$ be the marginal of $m(s_1, \pi_1, \pi_2)$ on $S_t \times A_t^1 \times A_t^2$. The tth stage reward of Pi is

$$r_i^{(t)}(\pi_1, \pi_2)(s_1) = \int r_i(s_t, a_t^1, a_t^2) dm_t(s_1, \pi_1, \pi_2), \quad i = 1, 2.$$

It is easily seen that

$$I_i(\pi_1, \pi_2)(s_1) = \sum_{t=1}^{\infty} \beta^{t-1} r_i^{(t)}(\pi_1, \pi_2)(s_1), \quad i = 1, 2.$$

A pair of Markovian strategies (π_1^*, π_2^*) is a Nash equilibrium if $I_1(\pi_1^*, \pi_2^*)(s_1) \geq I_1(\pi_1, \pi_2^*)(s_1)$ and $I_2(\pi_1^*, \pi_2^*)(s_1) \geq I_2(\pi_1^*, \pi_2)(s_1)$ for all Markovian strategies π_1 and π_2 and all initial states s_1.

If p is a probability measure on S with $q(\bullet \mid s, a_1, a_2) \ll p$ ($\ll$ denotes absolute continuity of measures), for every $(s, a_1, a_2) \in S \times A_1 \times A_2$, a p-equilibrium of the stochastic game is defined by the same inequalities, but only for p-almost every initial state s_1.

Our main results are:

Theorem I: Let A_i be a compact metrizable space, $i = 1, 2$ and the horizon be finite. The stochastic game has a Nash equilibrium in Markovian strategies.

The following provides an alternative proof of the result in Whitt [15] and Nowak [8].

Corollary Ia: Let A_i be a compact metrizable space and the horizon be infinite. The stochastic game has an ϵ-Nash equilibrium in Markovian strategies.[1]

The following assumption of dominated transitions is standard in the literature, and is needed for Theorem II:

D: There exists a probability measure p on S such that $q(\bullet \mid s, a_1, a_2) \ll p$ for all $(s, a_1, a_2) \in S \times A_1 \times A_2$.

It is interesting to note that D is implied by the assumption

C: $q(E \mid s, a_1, a_2)$ is jointly continuous on $S \times A_1 \times A_2$, for every Borel set $E \subseteq S$

as we now prove. C has been assumed in some previous investigations, e.g., Whitt [15], Parthasarathy [10], ...

[1]A. Nowak pointed out to the author that versions of Theorem I and Corollary Ia were obtained by U. Rieder in "Equilibrium Plans for Nonzero-sum Markov Games", in Game Theory and Related Topics, Moeschlin and Pallaschke, eds., North-Holland, Amsterdam [1979].

Proposition: Assumption C implies Assumption D.

Proof: Let $\{s_i\}$, $\{a_i^1\}$ and $\{a_i^2\}$ be countable dense sets in the spaces S, A_1 and A_2, respectively, all of which are clearly separable. Define a probability measure p on S by

$$p(E) = \sum_{s_i} \sum_{a_k^1} \sum_{a_j^2} C(i,j,k) q(E \mid s_i, a_k^1, a_j^2), \text{ for every Borel set } E,$$

where the constants $C(i,j,k)$ are positive and chosen so that $p(S) = 1$. Now, if $p(E) = 0$, it follows that $q(E \mid s_i, a_k^1, a_j^2) = 0$ for all s_i, a_k^1, a_j^2, and hence $q(E \mid s, a_1, a_2) = 0$ for all $(s, a_1, a_2) \in S \times A_1 \times A_2$, as a consequence of Assumption C. ◇

Theorem II: Let A_i be a finite set, $i = 1,2$ and the horizon be infinite. The stochastic game has a Nash equilibrium in Markovian strategies.

Proofs are given in the next section. Use is made of the following:

Fixed-Point Theorem: (Glicksberg [6]):

Let C_i be a compact convex subset of a locally convex Hausdorff topological vector space, $i = 1,2$, and $f_i : C_1 \times C_2 \to R$ be jointly continuous, and quasiconcave in each variable. Then there exists $c_i^* \in C_i$, $i = 1,2$, such that

$$f_1(c_1^*, c_2^*) \geq f_1(c_1, c_2^*), \qquad \forall c_1 \in C_1$$
$$f_2(c_1^*, c_2^*) \geq f_2(c_1^*, c_2), \qquad \forall c_2 \in C_2.$$

Before providing a discussion of our results, we note that the major difficulty in dealing with such stochastic games rests on the fact that payoffs are not jointly continuous in the strategies in the topology $\sigma\{L_\infty(S; p; M(A_i)), L_1(S, p; C(A_i))\}$, which is the weak-star topology on the space of all p-essentially bounded Borel-measurable functions from S into $M(A_i)$, the Banach space of bounded regular measures on A_i with variation norm, itself the dual of the Banach space $C(A_i)$ of all continuous functions on A_i with sup norm (cf. Diestel and Uhl [3]). As our strategy spaces lie on the unit ball of the above L_∞ space, this weak-star topology (is the only one that) guarantees their compactness. This difficulty has been encountered in similar contexts, e.g. Warga [14].

Finally, we note that the derived equilibria are subgame-perfect in the strong sense of the principle of optimality of dynamic programming. We remind the reader that Markovian Nash equilibria remain equilibria when players are allowed to use history dependent strategies.

Finally, it is worth noting that the equilibrium in Theorem II is arrived at as a limit of finite-horizon Markovian equilibria (from Theorem I), in an appropriate sense (cf. Lemma 3.4 in the next section). Also, to prove Theorem II, we first prove existence of a p-equilibrium and then, using a construction due to Parthasarathy [10], we construct an equilibrium.

3. Proofs:

The proofs are based on a series of preliminary results. For a given metric space (M, d), let $B(M)$ be the Banach space of bounded Borel measurable functions on M with sup norm, $\mathcal{B}(M)$ be the Borel σ-algebra of M, $P(M)$ be the set of all probability measures on M, and $P_S(M)$ be the set of all Borel maps from S to $P(M)(P(M)$ being endowed with the weak topology). Let $f_i : S \times (A_1 \times A_2) \to R$, $i = 1, 2$, be a uniformly bounded Caratheodary function, i.e. $f_i(s, \bullet, \bullet)$ is continuous on $A_1 \times A_2$ for each $s \in S$ and $f(\bullet, a_1, a_2)$ is Borel measurable on S for each $(a_1, a_2) \in A_1 \times A_2$.

Lemma 3.1: For every fixed $s \in S$, the one-shot game with payoffs

$$f_i(s, \mu_1, \mu_2) = \int_{A_1} \int_{A_2} f_i(s, a_1, a_2)\mu_1(da_1)\mu_2(da_2)$$

has a Nash equilibrium in $P(A_1) \times P(A_2)$.

Proof: $P(A_i)$, $i = 1, 2$, endowed with the weak-star topology as a subset of the dual of the Banach space of continuous functions on A_i with sup norm $C(A_i)$ (alias weak topology in probability theory), is compact. f_i is clearly jointly continuous on $P(A_i) \times P(A_2)$, and affine on $P(A_1)$ and $P(A_2)$. By Glicksberg's fixed-point theorem, a Nash equilibrium exists. ◇

Corollary 3.2: For every fixed $s \in S$ and $v_i \in B(i)$, $i = 1, 2$, the one-shot game with payoffs

$$h_i(s, \mu_1, \mu_2) = r_i(s, \mu_1, \mu_2) + \int_S v_i(s')q(ds'; s, \mu_1, \mu_2), \quad i = 1, 2,$$

has a Nash equilibrium in $P(A_1) \times P(A_2)$.

Proof: By Proposition 11.18 of Royden [11], the term $\int_S v_i(s')q(ds'; s, \mu_1, \mu_2)$ is sequentially continuous on $P(A_1) \times P(A_2)$. Since $P(A_i)$, $i = 1, 2$, with the weak topology, is metric, it follows that h_i has all the properties of f_i (from Lemma 3.1). The conclusion now derives from Lemma 3.1. ◇

We now show that as s varies in S, a pair of Borel measurable functions may be selected out of the Nash equilibrium strategy correspondence. This generalizes the selection theorem in Parthasarathy [10] where joint continuity over $S \times A_1 \times A_2$ is assumed.

Lemma 3.3: There exist two Borel measurable functions μ_1^* and $\mu_2^*, \mu_i^* : S \to P(A_i)$, $i = 1, 2$, such that

$$h_1(s, \mu_1^*(s), \mu_2^*(s)) \geq h_1(s, \mu_1, \mu_2^*(s)), \qquad \text{for all } s \in S,\ \mu_1 \in P(A_i),$$
$$h_2(s, \mu_1^*(s), \mu_2^*(s)) \geq h_2(s, \mu_1^*(s), \mu_2(s)), \qquad \text{for all } s \in S,\ \mu_2 \in P(A_2).$$

Proof: Define the set-valued map:

$$\varphi : S \to 2^{P(A_1) \times P(A_2)}$$
$$s \to \varphi(s) = \{(\lambda_1, \lambda_2) : h_1(s, \lambda_1, \lambda_2) \geq h_1(s, \mu_1, \lambda_2),\ \forall \mu_1 \in P(A_1)$$
$$\text{and } h_2(s, \lambda_1, \lambda_2) \geq h_2(s, \lambda_1, \mu_2),\ \forall \mu_2 \in P(A_2)\}.$$

First, note that h_i is $\mathcal{B}(S) \oplus \mathcal{B}[P(A_1)] \oplus \mathcal{B}[P(A_2)]$ measurable. $P(A_i)$, being compact metric in the weak topology, is separable. Let $(\mu_i)_{k=1}^{\infty}$ be a countable dense subset, $i = 1, 2$. Then, since h_i is continuous on $P(A_1) \times P(A_2)$, if K is a closed subset of $P(A_1) \times P(A_2)$, we have

$$\varphi^{-1}(K) = \{s \in S : h_1(s, \lambda_1, \lambda_2) \geq h(s, \mu_1^k, \lambda_2) \text{ and}$$
$$h_2(s, \lambda_1, \lambda_2) \geq h_2(s, \lambda_1, \mu_2^k), k = 1, 2, \ldots, (\lambda_1, \lambda_2) \in K\}$$
$$= \bigcap_{k=1}^{\infty} \{(s \in S) : h_1(s, \lambda_1, \lambda_2) \geq h_1(s, \mu_1^k, \lambda_2) \text{ and}$$
$$h_2(s, \lambda_1, \lambda_2) \geq h_2(s, \lambda_1, \mu_2^k), (\lambda_1, \lambda_2) \in K\}.$$

Hence $\varphi^{-1}(K) \in \mathcal{B}(S)$, and φ is a measurable multifunction. φ is also compact valued. By Theorem 4.2 in Wagner [13], there exists a measurable selection (μ_1^*, μ_2^*) out of φ. ◇

Proof of Theorem I:

The one-period horizon game has payoffs $r_i(s, \mu_1, \mu_2)$. By Lemma 3.3, it has a Nash equilibrium. Let $(v_1^1(s), v_2^1(s))$ be corresponding (Borel measurable) equilibrium return functions. By Corollary 3.2 (for Player i, $i = 1, 2$) and Lemma 3.3, the two-period horizon game has a Nash equilibrium $(\mu_1^2(s), \mu_2^2(s))$ with corresponding value functions $(v_1^2(s), v_2^2(s))$.

The conclusion follows from iterated use of the above argument. ◇

Proof of Corollary Ia:
Given $\epsilon > 0$, choose N sufficiently large so that $\sum_{t=N+1}^{\infty} \beta^t K < \epsilon$, or equivalently $N \geq log\,\beta^{-1}(\frac{K}{\epsilon(1-\beta)}) - 2$, where K is the upper bound on the one-period reward functions. If Player i uses his Nash equilibrium Markovian strategy $(\mu_i)_{t=1}^{N}$ up to stage N and then follows any feasible strategy for the remainder of the game, $i = 1, 2$, the resulting payoffs clearly form an ϵ-equilibrium. ◊

We now proceed to the proof of Theorem II via intermediate steps. Further notation is needed. Define a multi-valued operator $T : B(S) \times B(S) \to 2^{B(S)\times B(S)}$: $(v_1, v_2) \to \{(Tv_1, Tv_2)\}$, where $\forall s \in S$, $\forall \lambda_i \in P(A_i)$, $i = 1, 2$:

$$\begin{aligned} Tv_1(s) &= r_1(s, \mu_1^*(s), \mu_2^*(s)) + \beta_1 \int_S v_1(s')q(ds'; s, \mu_1^*(s), \mu_2^*(s)) \\ &\geq r_1(s, \lambda_1, \mu_2^*(s)) + \beta_1 \int_S v_1(s')q(ds'; s, \lambda_1, \mu_2^*(s)) \end{aligned}$$

and

$$\begin{aligned} Tv_2(s) &= r_2(s, \mu_1^*(s), \mu_2^*(s)) + \beta_2 \int_S v_2(s')q(ds'; s, \mu_1^*(s), \mu_2^*(s)) \\ &\geq r_2(s, \mu_1^*(s), \lambda_2) + \beta_2 \int_S v_2(s')q(ds'; s, \mu_1^*(s), \lambda_2) \end{aligned}$$

for a measurable selection (μ_1^*, μ_2^*).

Let $L_\infty(p)$ denote the Banach space of (equivalence classes of) p-essentially bounded real-valued measurable functions on S with ess-sup norm. Weak-star convergence in $L_\infty(p)$ will be denoted by $\xrightarrow{w^*}$, while strong or weak convergence of measures in $P(A_i)$, $i = 1, 2$, by $\to$. Given a sequence (f_n) in $L_\infty(p)$, let $L(f_n(s))$ denote the set of limit points of the sequence $(f_n(s))$ in R.

Finally, $L_1(p)$, the predual of $L_\infty(p)$, is the Banach space of p-integrable functions from S to $\mathbf{R}$ with norm $\int_S |f|dp$.

The following result is referred to as a generalized Fatou lemma.

Lemma 3.4: Let $\{f_n\}$ be a uniformly integrable sequence in $L_1(p)$ such that $lim_n \int f_n dp$ exists. Then there exists $\hat{f} \in L_1(p)$ with $\hat{f}(s) \in L(f_n(s))$ for p-almost all $s \in S$, and $\int_S \hat{f} dp = lim_n \int_S f_n dp$.

Proof: This is Corollary 1 in Balder [1], with $m = 1$. ◊

Lemma 3.5: For a fixed $s \in S$, let $\mu_n^i(s) \to \mu_i(s)$ in $P(A_i)$, $i = 1, 2$, and $v_n \xrightarrow{w^*} v$ in $L_\infty(p)$. Then

$$\int_S v_n(s')q(ds'; s, \mu_n^1(s), \mu_n^2(s)) \to \int_S v_n(s')q(ds'; s, \mu_1(s), \mu_2(s)).$$

Proof: Let

$$\varphi_n(s, a_1, a_2) = \int_S v_n(s')q(ds'; s, a_1, a_2) \text{ and}$$

$$\varphi(s, a_1, a_2) = \int_S v(s')q(ds'; s, a_1, a_2), \ a_i \in A_i, \ i = 1, 2.$$

Since $q(\bullet; s, a_1, a_2) \ll p$ and $v_n \xrightarrow{w^*} v$, we have $\forall \ a_i \in A_i$, $i = 1, 2$, and fixed $s \in S$, $\varphi_n(s, a_1, a_2) \to \varphi(s, a_1, a_2)$. Now, one easily sees that

$$\begin{aligned} \int_S v(s')q(ds'; s, \mu_n^1(s), \mu_n^2(s)) &= \sum_{a_1}\sum_{a_2} \varphi_n(s, a_1, a_2)\mu_n^1(a_1; s)\mu_n^2(a_2; s) \\ &\to \sum_{a_1}\sum_{a_2} \varphi(s, a_1, a_2)\mu_1(a_1; s)\mu_2(a_2; s) \\ &= \int v(s')q(ds'; s, \mu_1(s), \mu(s)). \end{aligned}$$

◇

We are now ready for

Proof of Theorem II:

Let $(v_n^1, v_n^2) \in L_\infty(p) \times L_\infty(p)$ be equilibrium value selections in the n-period horizon stochastic game (cf. Theorem I), $n \in \mathbf{N}$, with associated first period decision $(\mu_n^1, \mu_n^2) \in P_S(A_1) \times P_S(A_2)$ (see definition of the operator T). Since $\{v_n^i\}$ is uniformly bounded, there exists $M_1 \subset \mathbf{N}$ such that $v_{m_1}^i \xrightarrow{w^*} u_1^i$, $M_2 \subset M_1$ such that $v_{(m_2-1)}^i \xrightarrow{w^*} u_2^i$, $M_3 \subset M_2$ such that $v_{(m_3-2)}^i \xrightarrow{w^*} u_3^i, \ldots, v_{(m_k-k+1)}^i \xrightarrow{w^*} u_k^i, \ldots, i = 1, 2$. This process generates a uniformly bounded sequence $\{u_k^1, u_k^2\}$ in $L_\infty(p) \times L_\infty(p)$.

Now, since $(v_n^1, v_n^2) \in T(v_{n-1}^1, v_{n-1}^2)$, we also have

$$(v_{(m_k-k+1)}^1, v_{(m_k-k+1)}^2) \in T(v_{m_k-k}^1, v_{m_k-k}^2), \quad k = 1, 2, 3, \ldots$$

While the first pair (on LHS) of sequences converges w^* according to our construction, the second pair need not. To ensure that both pairs converge w^*, we instead consider the subsequence

$$(v^1_{(m_{k+1}-k+1)}, v^2_{(m_{k+1}-k+1)}) \in T(v^1_{(m_{k+1}-k)}, v^2_{(m_{k+1}-k)}), \quad k = 1, 2, 3, \ldots \quad (1)$$

where

$$v^i_{(m_{k+1}-k+1)} \xrightarrow{w^*} u^i_k \text{ and } v^i_{(m_{k+1}-k)} \xrightarrow{w^*} u^i_{k+1}, \quad i = 1, 2. \quad (2)$$

Applying Lemma 3.4 to $\{v^i_{m_2}\}$, we conclude that there exists $\hat{u}^i_1 \in L_\infty(p)$ such that

$$\hat{u}^i_1(s') \in L\{v^i_{m_2}(s')\}p \text{ a.e. and } \int_S \hat{u}^i_1(s')dp = \lim_{m_2} \int_S v^i_{m_2}(s')dp = \int_S \hat{u}_1(s')dp.$$

Considering (1) with $k = 1$, we have, say for Player I (where $h = dq/dp$):

$$\begin{aligned} v^1_{m_2}(s) &= r(s, \mu^1_{m_2}(s), \mu^2_{m_2}(s)) + \beta_1 \int_S v^1_{(m_2-1)}(s')q(ds'; s, \mu^1_{m_2}(s), \mu^2_{m_2}(s)) \\ &= r(s, \mu^1_{m_2}(s), \mu^2_{m_2}(s)) + \beta_1 \int_S v^1_{(m_2-1)}(s')h(s'; s, \mu^1_{m_2}(s), \mu^2_{m_2}(s))dp \\ &\geq r(s, \lambda, \mu^2_{m_2}(s)) + \beta_1 \int_S v^1_{(m_2-1)}(s')h(s'; s, \lambda, \mu^2_{m_2}(s))dp, \end{aligned} \quad (3)$$

for every $s \in S$, $\lambda \in P_{A_1}$.

Now, *for every* $s \in S$, take the subsequence of $\{v^1_{m_2}(s)\}$, denoted $\{v_\ell(s)\}$, which converges to $\hat{u}^1_1(s)$ in $\mathbf{R}$. By Lemma 3.5, the corresponding subsequence of the RHS converges in $\mathbf{R}$ to

$$r(s, \mu^1_1(s), \mu^2_1(s)) + \beta_1 \int_S u^1_2(s')h(s'; s, \mu^1_1(s), \mu^2_1(s))dp.$$

Clearly, the sequence $\{v^1_{(\ell-1)}(s')h(s'; s, \mu^1_\ell(s), \mu^2_\ell(s))\}$ is uniformly integrable in $L_1(p)$, for each $s \in S$, and moreover

$$L\{v^1_{(\ell-1)}(s')h(s'; s, \mu^1_\ell(s), \mu^2_\ell(s))\} = h(s'; s, \mu^1_1(s), \mu^2_1(s))L\{v^1_{(\ell-1)}(s')\}.$$

Hence, by Lemma 3.4, there exists $\hat{u}^1_2 \in L_\infty(p)$ such that $\hat{u}^1_2(s') \in L\{v^1_{(\ell-1)}(s')\}p$ a.e. and

$$\int_S \hat{u}^1_2(s')h(s'; s, \mu^1_1(s), \mu^2_1(s))dp = \int_S u^1_2(s')h(s'; s, \mu^1_1(s), \mu^2_1(s))dp.$$

Therefore, we conclude

$$(\hat{u}_1^1, \hat{u}_1^2) \in T(\hat{u}_2^1, \hat{u}_2^2) = T(u_2^1, u_2^2) \text{ with } (\mu_1^1, \mu_1^2) \in P_S(A_1) \times P_S(A_2)$$

as first-period equilibrium decision functions, i.e.

$$\begin{aligned}\hat{u}_1^1(s) &= r(s, \mu_1^1(s), \mu_1^2(s)) + \beta_1 \int_S u_2^1(s')q(ds; s, \mu_1^1(s), \mu_1^2(s)) \\ &= r(s, \mu_1^1(s), \mu_1^2(s)) + \beta_1 \int_S \hat{u}_2^1(s')q(ds'; s, \mu_1^1(s), \mu_1^2(s)) \\ &\geq r(s, \lambda, \mu_1^2(s)) + \beta_1 \int_S \hat{u}_2^1(s')q(ds'; s, \lambda, \mu_1^2(s)),\end{aligned}$$

for p almost all $s \in S$ and every $\lambda \in P_{A_1}$.

Starting from (1) with $k = 2$, repeat the same argument for fixed $s \in S$ with the new index ℓ determined by convergence of $\{v^1_{(m_s-1)}(s)\}$ to $\hat{u}_2^1(s)$ in $\mathbf{R}$... Iterate for $k = 3, 4, \ldots$ In conclusion, we get

$$(\hat{u}_k^1, \hat{u}_k^2) \in T(u_{k+1}^1, u_{k+1}^2) = T(\hat{u}_{k+1}^1, \hat{u}_{k+1}^2), \quad k = 1, 2, 3, \ldots, \tag{4}$$

with $(\mu_k^1, \mu_k^2) \in P_S(A_1) \times P_S(A_2)$ as first-period equilibrium decision functions.

We now show that playing $\mu_k^i(\bullet)$ at stage k, for Pi, $i = 1, 2$, $k = 1, 2, \ldots$, constitutes the claimed Nash equilibrium, with $\hat{u}_k^i$ as payoff from stage k onwards. To this end, consider the augmented state space $S \times N$, and the sequence $\{\mu_k^i(\bullet)\}$ as "stationary" strategies with respect to this new space. The conclusion follows by regarding equations (4) as the optimality equations for the discounted stationary dynamic programming problems faced by Players I and II, respectively when $\{u_k^2\}$ and $\{\mu_k^1\}$ are fixed (Cf. Proposition 10.1 in Bertsekas-Shreve [2]).

This completes the proof of existence of a p-equilibrium. To establish the existence of an equilibrium, we appeal to the construction suggested by Parthasarathy [10] together with our more general measurable selection result (Lemma 3.3).

References

[1] *Balder, E. J.* (1984). A Unifying Note on Fatou's Lemma in Several Dimensions, *Math. of Oper. Res.*, **9**, 267-275.

[2] *Bertsekas, D. and S. Shreve* (1978). "Stochastic Optimal Control: The Discrete-Time Case", Academic Press.

[3] *Diestel, J. and J. Uhl* (1977). "Vector Measures", Math. Surveys, vol. 15, A.M.S., Providence.

[4] *Dunford, N. and J. Schwartz* (1958). "Linear Operators, Part I", Interscience, New York.

[5] *Federgruen, A.* (1978). On *N*-Person Stochastic Games with Denumerable State Space, *Adv. in Appl. Probability*, **10**, 452-471.

[6] *Glicksberg, I.* (1952). A Further Generalization of Kakutani's Fixed Point Theorem with Applications to Nash Equilibrium Points, *Proc. Nat. Acad. Sci., USA*, **38**, 170-172.

[7] *Himmelberg, C., T. Parthasarathy, T. Raghavan and F. Van Vleck* (1976). Existence of *p*-Equilibrium and Optimal Stationary Strategies in Stochastic Games, *Proc. Amer. Math. Soc.*, **60**, 245-251.

[8] *Nowak, S.* (1985). Existence of Equilibrium stationary Strategies in Discounted Noncooperative Stochastic Games with Uncountable State Space, *J. Opt. Theory Appl.*, **45**, 591-602.

[9] *Parthasarathy, T.* (1973). Discounted, Positive and Noncooperative Stochastic Games, *Int. J. Game Theory*, **1**, 25-37.

[10] *Parthasarathy, T.* (1982). Existence of Equilibrium Stationary Strategies in Discounted Stochastic Games, *Sankhya*, **44**, 114-127.

[11] *Royden, H.* (1968). "Real Analysis", MacMillan, New York.

[12] *Shapley, L.* (1953). Stochastic Games, *Proc. of Nat. Acad. of Sci., USA*, **39**, 1095-1100.

[13] *Wagner, D.* (1977). Survey of Measurable Selection Theorems, *SIAM J. on Cont. and Opt.*, **15**, 859-1003.

[14] *Warga, J.* (1972). Optimal Control of Differential and Functional Equations, Academic Press, New York.

[15] *Whitt, W.* (1980). Representation and Approximation of Noncooperative Sequential games, *SIAM J. on Cont. and Opt.*, **18**, 33-48.

Rabah Amir
Department of Economics
State University of New York at Stony Brook
Stony Brook, NY 11794
U.S.A.

Non-Cooperative Dynamic Games with General Utility Functions

Ulrich Rieder

1. Introduction and Summary: The present paper is concerned with a general non-cooperative two-person dynamic game with Borel state and action spaces, non-Markovian transition law and with utility functions depending on the whole sequence of states and actions. The motivation for a general utility function is that in several problems in economic theory, additivity or separability of the utility function is a restrictive assumption and hard to justify, e.g. in problems of consumption and production choices over time and in the closely related problems of optimal economic growth. Dynamic games with additive utility functions have been introduced by Shapley [22] and have then been investigated by many authors (see the survey paper of Parthasarathy and Stern [16] or Küenle [9]). In recent years several authors have considered dynamic games with more general utility functions, e.g. Sengupta [21], Iwamoto [7], Schäl [19].

The main purpose of the present paper is to give sufficient conditions for the existence of a Nash equilibrium (in the nonzero-sum game) and of a Minimax policy (in the zero-sum game). In the zero-sum case we are especially interested in "asymmetric" sufficient conditions. Our methods of proof emanate from non-stationary dynamic programming and utilize some results of semi-continuous functions and measurable selection theorems.

In section 2 we formulate the non-cooperative decision model. Nonzero-sum dynamic games with bounded utility functions are considered in section 3. Theorem 3.1 extends the related well-known result for discounted dynamic games (cf. also Nowak [13]), whereas Theorem 3.2 is the non-stationary version of results in Rieder [18] (cf. also Kindler [8] and Federgruen [3]). The second part of this paper contains the main results for zero-sum dynamic games which are stronger and more complete as in the nonzero-sum case. Here the utility function is only assumed to be bounded from below. We introduce the notion of a lower approximable utility function and present three different sets of Conditions (W), (S1) and (S2). Then it is shown that the zero-sum dynamic game is strongly determined and that there exists a Minimax policy for the second player. Moreover, the sequence of value functions is a solution of two functional equations - the so-called upper and lower optimality equations. By interchanging the role of the two players one obtains sufficient conditions for the existence of a Minimax policy for the first player. The results may be regarded as generalizations of results given by Sengupta [21], Iwamoto [7], Kindler [8], and Nowak [12], [14]. In particular, it turns out that the results by Maitra and Parthasarathy [10], [11], Parthasarathy [15], Frid [4] proved for the standard case of an additive utility function carry over to dynamic games with a lower approximable utility function (cf. also Rieder [17]).

Throughout this paper, the following notations will be used: $\mathbf{N}$:= the set of positive integers; $\mathbf{R}$:= the set of real numbers; $c^{\pm} := max(0, \pm c)$ for $c \in [-\infty, \infty]$; sequences of elements $a_1, a_2, \ldots$ are denoted by (a_n); for any measurable space $(X, \mathcal{A})$, let $P(X)$ stand for the set of all probability measures on $\mathcal{A}$;

$$\|v\| := \sup_x |v(x)| \qquad \text{for } v : X \to [-\infty, \infty].$$

2. The non-cooperative decision model: The mathematical framework for a non-cooperative dynamic game problem with two players consists of a tuple $\Gamma = ((S_n), (A_n), (B_n), (q_n), u_1, u_2)$ of objects of the following meaning:

(i) S_n is the *state space* at time n. S_n is assumed to be a Borel space, i.e. S_n is a non-empty Borel subset of some Polish (complete, separable metric) space and is endowed with the σ-algebra of Borel subsets of S_n.

(ii) A_n and B_n are the *action spaces* for player I and II, resp. A_n and B_n are assumed to be Borel spaces.

We write $H_n = S_n \times A_1 \times B_1 \times S_2 \times \ldots \times S_n$ and $H_\infty = S_1 \times A_1 \times B_1 \times S_2 \times \ldots$

(iii) q_n is a transition probability (or kernel) from $H_n \times A_n \times B_n$ to S_{n+1}, the so-called *transition law* between n and $n+1$. $Q_n(h, a, b, \cdot)$ is the distribution of the state of the system at time $n+1$ if both players have experienced the history h up to time n and player I and II choose action a and b, resp., at time n.

(iv) u_1 and u_2 are measurable functions on H_∞, the so-called *utility functions* for player I and II, resp.

Remark: For convenience, the set of admissible actions at time n is assumed to be independent of the history. However, the results of this paper extend to the more general case where such sets are specified for any finite history.

For any separable metric space X, the set of all probability measures $P(X)$ is endowed with the weak* topology based on bounded continuous functions. Then $P(X)$ is separable and metrizable. Further, the σ-algebra of Borel subsets on $P(X)$ coincides with the smallest σ-algebra on $P(X)$ such that for each Borel subset of X the map $p \to p(B)$ is measurable. If X is a Borel space then $P(X)$ is a Borel space. If X is compact metric, then $P(X)$ is compact and metrizable.

As usual, a (randomized) *policy* $\pi = (f_n)$ for player I is defined as a sequence of measurable maps $f_n : H_n \to P(A_n)$. Let Δ denote the set of all policies for player I. Similarly, the set Σ of all policies for player II is defined. Given some initial distribution $p \in P(S_1)$, by the Theorem of C. Ioncscu Tulcca any policy $\pi = (f_n) \in \Delta$ and any policy $\sigma = (g_n) \in \Sigma$ define a probability measure $P_{\pi\sigma} = pf_1g_1q_1f_2g_2q_2\ldots$ on H_∞ endowed with the product σ-algebra and thus a random process $\chi = (\varsigma_1, \alpha_1, \beta_1, \varsigma_2, \ldots)$ (cf. Hinderer [6], p.80) where ς_n, α_n and β_n denote the projection from H_∞ onto S_n, A_n and B_n, respectively. Then $\chi_n = (\varsigma_1, \alpha_1, \beta_1, \ldots, \varsigma_n)$ describes the history at time n. In this paper we are concerned only with the conditional distributions $P_{\pi\sigma}(\cdot|\chi_n)$ which are given by f_k, g_k, q_k, $k = n, n+1, \ldots$ and are independent of p. Therefore we can dispense with an initial distribution.

If u_i is bounded from below then $u_i \cdot \chi$ is quasi-integrable with respect to $P_{\pi\sigma}(\cdot|\chi_n = h)$ and hence the following functions

$$V_{n\pi\sigma}(h) = E_{\pi\sigma}[u_1 \cdot \chi|\chi_n = h], \quad n \in \mathbf{N}, \quad h \in H_n$$
$$W_{n\pi\sigma}(h) = E_{\pi\sigma}[u_2 \cdot \chi|\chi_n = h], \quad n \in \mathbf{N}, \quad h \in H_n$$

are well-defined for all $\pi \in \Delta$, $\sigma \in \Sigma$. $V_{n\pi\sigma}(h)$ and $W_{n\pi\sigma}(h)$ are the *expected utilities for player I and II*, resp., if the history at time n is h and player I and II use the policies π and σ, resp.

3. Nonzero-sum dynamic games: Throughout this section we assume that the utility functions u_1 and u_2 are bounded. Let $\epsilon \geq 0$. Then $(\pi, \sigma) \in \Delta \times \Sigma$ is called a *strongly ϵ-equilibrium (in the sense of Nash)* if

$$V_{n\pi\sigma} \geq V_{n\delta\sigma} - \epsilon \text{ for all } \delta \in \Delta, \quad n \in \mathbf{N} \tag{3.1}$$
$$W_{n\pi\sigma} \geq W_{n\pi\tau} - \epsilon \text{ for all } \tau \in \Sigma, \quad n \in \mathbf{N}.$$

The pair $(\pi, \sigma) \in \Delta \times \Sigma$ is called a *strongly equilibrium* if it is a strongly 0-equilibrium. Sufficient conditions for the existence of equilibrium pairs are given in the following theorems.

Theorem 3.1: Let S_n be countable and let A_n and B_n be finite, $n \in \mathbf{N}$. If there exists a sequence (u_{ni}) of bounded functions $u_{ni} : H_n \to \mathbf{R}$ such that (u_{ni}) converges uniformly to u_i for $i = 1, 2$, then there exists a strongly equilibrium pair.

Proof: Let $\pi = (f_n) \in \Delta$ and $\sigma = (g_n) \in \Sigma$. Define

$$F_{n\sigma} = \sup_{\delta} V_{n\delta\sigma} \text{ and } G_{n\pi} = \sup_{\tau} W_{n\pi\tau}, (\delta, \tau) \in \Delta \times \Sigma.$$

Note that $F_{n\sigma}$ and $G_{n\pi}$ depend on σ and π only through g_k and f_k, $k = n, n+1, \ldots$ By means of a non-stationary dynamic program we conclude from the results in Hinderer [6] that

$$F_{n\sigma} = \sup_{f}(f g_n q_n F_{n+1,\sigma}), n \in \mathbf{N}$$

$$G_{n\pi} = \sup_{g}(f_n g q_n G_{n+1,\pi}), n \in \mathbf{N}$$

where the supremum is taken over the set of all functions $f : H_n \to P(A_n)$ and $g : H_n \to P(B_n)$, resp. Furthermore, since $\|u_{ni} - u_i\| \to 0$ $(n \to \infty)$, $i = 1, 2$, we get the following characterization: (π, σ) is a strongly equilibrium if and only if (3.2) and (3.3) are satisfied.

$$F_{n\sigma} = f_n g_n q_n F_{n+1,\sigma}, \quad n \in \mathbf{N} \tag{3.2}$$

$$G_{n\pi} = f_n g_n q_n G_{n+1,\pi}, \quad n \in \mathbf{N} \tag{3.3}$$

In view of the assumptions for S_n, A_n and B_n we may look upon Δ and Σ as compact convex subsets of some locally convex linear topological Hausdorff space. (For a proof use the topology of pointwise convergence.) Define for $\pi = (f_n) \in \Delta$ and $\sigma = (g_n) \in \Sigma$

$$\Phi_1(\pi, \sigma) = \{(f'_n) \in \Delta : f'_n g_n q_n F_{n+1,\sigma} = F_{n\sigma} \text{ for all } n \in \mathbf{N}\}$$

$$\Phi_2(\pi, \sigma) = \{(g'_n) \in \Sigma : f_n g'_n q_n G_{n+1,\pi} = G_{n\pi} \text{ for all } n \in \mathbf{N}\}$$

$$\Phi(\pi, \sigma) = \Phi_1(\pi, \sigma) \times \Phi_2(\pi, \sigma)$$

Obviously $\Phi(\pi, \sigma)$ is non-empty. Let $n \in \mathbf{N}$ and $h \in H_n$ be fixed. Then it is not difficult to establish that the functions

$$(\pi, \sigma) \to E_{\pi\sigma}[u_{mi} \cdot \chi_m | \chi_n = h], m \in \mathbf{N}, i = 1, 2$$

and thus

$$(\pi, \sigma) \to V_{n\pi\sigma}(h) \quad \text{and} \quad (\pi, \sigma) \to W_{n\pi\sigma}(h)$$

are continuous on $\Delta \times \Sigma$. Hence $\sigma \to F_{n\sigma}(h)$ and $\pi \to G_{n\pi}(h)$ are continuous. Moreover, the functions

$$(\pi, \sigma, (f'_n)) \to (f'_n g_n q_n F_{n+1,\sigma})(h)$$
$$(\pi, \sigma, (g'_n)) \to (f_n g'_n q_n G_{n+1,\pi})(h)$$

are continuous on $\Delta \times \Sigma \times \Delta$ and $\Delta \times \Sigma \times \Sigma$, respectively. It then follows that Φ is a closed correspondence of $\Delta \times \Sigma$ to $\Delta \times \Sigma$ with compact convex values. By the Glicksberg-Fan fixed point theorem ([5],[2]), there exists a pair $(\pi, \sigma) \in \Delta \times \Sigma$ such that $(\pi, \sigma) \in \Phi(\pi, \sigma)$. By (3.2) and (3.3), (π, σ) is a strongly equilibrium pair. $\diamond$

The extension of Theorem 3.1 to more general state and action spaces seems to be an open problem. However, a first result is valid if the underlying dynamic game Γ is *Markovian* and the utility functions u_1 and u_2 are *separable*, i.e. if q_n depends on the whole history $h \in H_n$ only through s_n and if

$$u_i(h_\infty) = \sum_{n=1}^{\infty} d_{1i}(s_1, a_1, b_1) \ldots d_{n-1,i}(s_{n-1}, a_{n-1}, b_{n-1}) r_{ni}(s_n, a_n, b_n)$$

where the functions $d_{ni} : S_n \times A_n \times B_n \to [0, \infty)$ and $r_{ni} : S_n \times A_n \times B_n \to \mathbf{R}$ are bounded and measurable.

Conditions (A):

(1) A_n and B_n are compact, $n \in \mathbf{N}$.

(2) $(a, b) \to q_n(h, a, b, B)$ is continuous for all Borel subsets B of S_{n+1} and $h \in H_n$, $n \in \mathbf{N}$.

(3) $d_{ni}(s, \cdot)$ and $r_{ni}(s, \cdot)$ are continuous on $A_n \times B_n$ for all $s \in S_n$, $i = 1, 2$, $n \in \mathbf{N}$.

If u_1 and u_2 depend only on the first $3N-2$ coordinates $(s_1, a_1, b_1, s_2, \ldots, s_N) \in H_N$, then the game problem of section 2 can be regarded as a N-stage dynamic game or as a decision model with finite horizon N. In this case we need only consider the first $N - 1$ components of a policy.

Theorem 3.2: Let the dynamic game be Markovian and let the utility functions u_1 and u_2 be separable. Let the conditions (A) be satisfied.

(a) If u_1 and u_2 depend only on the first $3N - 2$ coordinates then there exists a strongly equilibrium pair (of Markovian policies).

(b) Let $sup_n \|d_{ni}\| < 1$ and $sup_n \|r_{ni}\| < \infty$, $i = 1, 2$. Then for every $\epsilon > 0$, there exists a strongly ϵ-equilibrium pair (of Markovian policies), and, if moreover S_n is countable for $n \in \mathbf{N}$, there exists a strongly equilibrium pair (of Markovian policies).

Proof: A proof of (a) follows by induction where use is made of a measurable selection theorem (cf. section 4.1). The first statement of (b) is easily obtained from (a). The last assertion can be proven in the same way as Theorem 3.1. On the other hand, since the non-stationary Markovian dynamic game can be reformulated as a stationary model with a transition law given by a bounded transition measure, the statements of Theorem 3.2 can also be derived directly via the results for stationary models (cf. Rieder [18]). ◇

We remark that Theorems 3.1 and 3.2 can be extended to non-cooperative dynamic games with a countable set of players.

4. Zero-sum dynamic games: In the remainder of this paper we will consider zero-sum dynamic games. We set

$$u := u_1 = -u_2$$

and we assume that u is bounded from below. Define

$$V_n = \inf_{\sigma} \sup_{\pi} V_{n\pi\sigma}, n \in \mathbf{N}. \tag{4.1}$$

The map $h \to V_n(h)$ is called the *upper value function* at time n. The zero-sum dynamic game Γ is *strongly determined*, if

$$\sup_{\pi} \inf_{\sigma} V_{n\pi\sigma} = \inf_{\sigma} \sup_{\pi} V_{n\pi\sigma}, n \in \mathbf{N}$$

and in this case V_n is called the value function at time n. If Γ is strongly determined then $\pi \in \Delta$ will be called *strongly optimal for player I* if $\inf_\sigma V_{n\pi\sigma} = V_n$ for all $n \in \mathbf{N}$, and $\sigma \in \Sigma$ will be called *strongly optimal for player II* if $\sup_\pi V_{n\pi\sigma} = V_n$ for all $n \in \mathbf{N}$. In the next section we present sufficient conditions for the existence of strongly optimal policies.

Formulations simplify considerably by the use of the following notation. Let X be an arbitrary set. For any sequence of functions $v_n : X \to (-\infty, \infty]$ we write $v_n \nearrow$ if

$$\sup_{n \geq m} \sup_{x} (v_m(x) - v_n(x))^+ \to 0 \quad (m \to \infty)$$

where $\infty - \infty = 0$. The property $v_n \nearrow$ implies that $v = \lim_n v_n$ exists and we write $v_n \nearrow v$. If the functions v_n are finite, then the sequence (v_n) converges uniformly if and only if $v_n \nearrow$ and $(-v_n) \nearrow$.

We need some properties of semi-continuous functions. The following result is Theorem 3.1 in Rieder [17].

Lemma 4.1: Let A and B be metric spaces. Let (v_n) be a sequence of functions $v_n : A \times B \to (-\infty, \infty]$ which are bounded from below and for which $v_n(a, \cdot)$ is lower semi-continuous on B, $a \in A$. Let $v_n \nearrow v$.

(a) $v(a, \cdot)$ is lower semi-continuous on B, $a \in A$.

(b) If B is compact then

$$\lim_n \inf_B \sup_A v_n = \inf_B \sup_A \lim_n v_n$$
$$\lim_n \sup_A \inf_B v_n = \sup_A \inf_B \lim_n v_n.$$

4.1. Selection theorems: The proof of the existence of optimal policies rests on selection theorems which are proved in Rieder [17] or Nowak [14] and which will be listed below. Throughout this subsection let X, A and B be Borel spaces. For any measurable functions $v : X \times A \times B \to (-\infty, \infty]$ which is bounded from below, the following functions

$$\overline{v}(x, \mu, \nu) = \int \mu(da) \int \nu(db) v(x, a, b) \quad x \in X, \mu \in P(A), \nu \in P(B)$$
$$v^*(x) = \inf_\nu \sup_\mu \overline{v}(x, \mu, \nu) \quad x \in X$$

are well-defined. In the sequel we will simply write v instead of $\overline{v}$.

Theorem 4.2: Let $v : X \times A \times B \to (-\infty, \infty]$ be measurable and bounded from below and let $v(x, \cdot)$ be lower semi-continuous on $A \times B$, $x \in X$. Let B be compact.

(a) v^* is measurable and

$$\sup_\mu \inf_\nu v(x, \mu, \nu) = \inf_\nu \sup_\mu v(x, \mu, \nu), \quad x \in X.$$

(b) There exists a measurable map $g : X \to P(B)$ such that

$$\sup_\mu v(x, \mu, g(x)) = v^*(x), x \in X.$$

Theorem 4.3: Let $v : X \times A \times B \to \mathbf{R}$ be measurable and bounded. Let $v(x, \cdot, b)$ be upper semi-continuous on A for $x \in X$, $b \in B$ and $v(x, a, \cdot)$ be lower semi-continuous on B for $x \in X$, $a \in A$. Let A and B be compact.

(a) v^* is measurable and

$$\sup_\mu \inf_\nu v(x,\mu,\nu) = \inf_\nu \sup_\mu v(x,\mu,\nu), x \in X.$$

(b) There exist measurable maps $f : X \to P(A)$ and $g : X \to P(B)$ such that

$$\begin{aligned}\sup_\mu v(x,\mu,g(x)) &= v^*(x)\\ &= \inf_\nu v(x,f(x),\nu), x \in X.\end{aligned}$$

5. Existence of optimal policies: As in the standard case of an additive utility we have to impose some convergence conditions on the utility function. Let $M(H_n)$ denote the set of all measurable functions $v : H_n \to (-\infty,\infty]$ which are bounded from below. The utility function u is called *lower approximable* if there exists a sequence (u_n) such that $u_n \in M(H_n)$ and $u_n \nearrow u$. (We agree to look at any function defined on H_n as a function defined on H_∞ which depends on the first $3N-2$ coordinates only.)

Examples: The function u is lower approximable if one of the following conditions is satisfied.

(1) $u = \sum_1^\infty r_n$ for some (r_n) such that $r_n \in M(H_{n+1})$ for all $n \in \mathbf{N}$ and $\sum_1^\infty \|r_n^-\| < \infty$.

(2) u is lower semi-continuous and S_n, A_n and B_n are compact spaces for all $n \in \mathbf{N}$. For a proof consider the lower semi-continuous function $h \to u_n(h) := \inf_x u(h,x)$.

This case has been studied by Sengupta [21].

(3) u depends on finitely many coordinates only.

Throughout this section we assume that u is lower approximable by some sequence (u_n). Now we will introduce the well-known *optimal reward operators*. The isotone operator U_n is well-defined on $M(H_{n+1})$ by

$$(U_n v)(h) := \inf_\nu \sup_\mu (q_n v)(h,\mu,\nu), n \in \mathbf{N}, h \in H_n \tag{5.1}$$

where $(q_n v)(h,a,b) := \int q_n(h,a,b,ds)v(h,a,b,s), (h,a,b) \in H_n \times A_n \times B_n$.

The aim of this section is to give sufficient conditions for the following statements provided that the expressions are defined:

Statement I: $V_n = \sup_\pi \inf_\sigma V_{n\pi\sigma}, \quad n \in \mathbf{N}$.

Statement II: $V_n = U_n V_{n+1}, \quad n \in \mathbf{N}$.

Statement III: $V_n = lim_m\, U_n \ldots U_{m-1} u_m, \quad n \in \mathbf{N}$.

Statement IV: There exists a strongly optimal policy for player II.

Statement I means that the dynamic game Γ is strongly determined. A sequence (v_n) of functions $v_n \in M(H_n)$ will be called a *solution of the upper optimality equation* if

$$v_n = U_n v_{n+1}, n \in \mathbf{N}. \tag{5.2}$$

Then statement II says that the sequence of upper value functions (v_n) is a solution of the upper optimality equation. Statement III implies that V_n may be calculated by value iteration. The following characterization of (V_n) among the set of all solutions of (5.2) can be proved as in the standard case of an additive utility.

Theorem 5.1: Let the statements II and III be satisfied.

(a) If $u_n \nearrow u$ then (V_n) is the smallest of those solutions (v_n) of (5.2) that satisfy $lim_n \|(u_n - v_n)^+\| = 0$.

(b) Let u be bounded. If (u_n) converges uniformly to u then (V_n) is the only solution (v_n) of (5.2) that satisfies $lim_n \|u_n - v_n\| = 0$.

We remark that also the policy iteration (for player II) is valid under the conditions (W), (S1) and (S2) below. The continuity and compactness conditions may be regarded as extensions of the corresponding conditions in dynamic programming problems (cf. e.g. Schäl [20]).

5.1. Conditions (W): As usual, the transition probability q_n is called *w-continuous* if
$(h, a, b) \to \int q_n(h, a, b, ds)v(s)$ is continuous for all continuous and bounded functions $v : S_{n+1} \to \mathbf{R}$. From the well-known Portmanteau-Theorem it follows that q_n is w-continuous for all open sets G of S_{n+1}. We write $C_0(H_n)$ for the set of all lower semi-continuous functions $v \in M(H_n)$. Throughout this subsection we impose:

Conditions (W):

(1) B_n is compact, $n \in \mathbf{N}$.

(2) q_n is w-continuous, $n \in \mathbf{N}$.

(3) $u_n \nearrow u$ for some (u_n) such that $u_n \in C_o(H_n)$, $n \in \mathbf{N}$.

Lemma 5.2: U_n is a map from $C_o(H_{n+1})$ into $C_o(H_n)$.

Proof: Let $v \in C_o(H_{n+1})$. By the theorem of Baire (cf. Ash [1], p. 390) there exists an increasing sequence (v_m) of bounded continuous functions v_m such that $v = sup_m\, v_m$. Then $q_n v = sup_m\, q_n v_m$. Since q_n is w-continuous it follows that $q_n v$ is lower semi-continuous on $H_n \times A_n \times B_n$. Moreover, $q_n v$ is lower semi-continuous on $H_n \times P(A_n) \times P(B_n)$. Now it is easy to see that $U_n v$ is lower semi-continuous. ◇

Theorem 5.3: If the conditions (W) are satisfied then $V_n \in C_o(H_n)$ for all n and the basic statements I, II, III and IV are valid.

Proof: By Lemma 5.2 the function V_n^m is well-defined by $V_n^m := U_n \ldots U_{m-1} u_m$, and $V_n^m \in C_o(H_n)$. For fixed n, $V_n^m \nearrow$ as $m \to \infty$ and thus $V_n^\infty := lim_m\, V_n^m$ exists. By Lemma 4.1(a), $V_n^\infty \in C_o(H_n)$. Applying the monotone convergence theorem we get

$$V_{n\pi\sigma} = \lim_m V_{n\pi\sigma}^m, n \in \mathbf{N} \tag{5.3}$$

where $V_{n\pi\sigma}^m = E_{\pi\sigma}[u_m \cdot \chi_m \mid \chi_n = \cdot]$, $\pi \in \Delta$, $\sigma \in \Sigma$. From Theorem 4.2 we conclude that

$$V_n^m = \sup_\pi \inf_\sigma V_{n\pi\sigma}^m$$

and hence by (5.3)

$$\begin{aligned} V_n^\infty &= \lim_m \sup_\pi \inf_\sigma V_{n\pi\sigma}^m = \sup_\pi \lim_m \inf_\sigma V_{n\pi\sigma}^m \\ &\leq \sup_\pi \inf_\sigma V_{n\pi\sigma} \leq V_n, n \in \mathbf{N}. \end{aligned} \tag{5.4}$$

The interchange of lim and sup is verified by Lemma 4.1. It follows from Lemma 4.1(b) that $V_n^\infty = U_n V_{n+1}^\infty$. By Theorem 4.2(b) there exists a measurable map $g_n : H_n \to P(B_n)$ such that

$$\sup_\mu (q_n V_{n+1}^\infty)(h, \mu, g_n(h)) = (U_n V_{n+1}^\infty)(h) = V_n^\infty(h), h \in H_n. \tag{5.5}$$

By induction we get for all measurable maps $f_k : H_k \to P(A_k)$

$$\begin{aligned} V_n^\infty &\geq f_n g_n q_n \ldots f_{m-1} g_{m-1} q_{m-1} V_m^\infty \\ &\geq f_n g_n q_n \ldots f_{m-1} g_{m-1} q_{m-1} u_m - \epsilon_m, m \in \mathbf{N} \end{aligned}$$

where $\epsilon_m := sup_{k \geq m}\, sup_x (u_m(x) - u_k(x))^+$. Letting $m \to \infty$ yields by (5.3) $V_n^\infty \geq V_{n\pi\sigma}$ for $\sigma := (g_1, g_2, \ldots)$ and for all $\pi := (f_1, f_2, \ldots)$ i.e.

$$V_n^\infty \geq \sup_\pi V_{n\pi\sigma} \geq V_n, n \in \mathbf{N}.$$

In view of (5.4) and (5.5) the proof is complete. ◇

We remark that for zero-sum Markovian games with separable utility function the basic statements can be established under weaker conditions (W).

5.2. Conditions (S1): The transition probability q_n will be called *s-continuous* if
$(a_1, b_1, \ldots, a_n, b_n) \to \int q_n(h, a_n, b_n, ds)v(s)$ is continuous for all measurable and bounded functions $v : S_{n+1} \to \mathbf{R}$. It is easily verified that q_n is s-continuous if and only if $(a_1, b_1, \ldots, a_n, b_n) \to q_n(h, a_n, b_n, B)$ is continuous for all Borel subsets B of S_{n+1}. Let $C_1(H_n)$ denote the set of all functions $v \in M(H_n)$ such that $(a_1, b_1, \ldots, a_{n-1} b_{n-1}) \to v(h_n)$ is lower semi-continuous. Throughout this subsection we impose

Conditions (S1):

(1) A_n is σ-compact and B_n is compact, $n \in \mathbf{N}$.

(2) q_n is s-continuous, $n \in \mathbf{N}$.

(3) $u_n \nearrow u$ for some (u_n) such that $u_n \in C_1(H_n)$, $n \in \mathbf{N}$.

Lemma 5.4: U_n is a map from $C_1(H_{n+1})$ into $C_1(H_n)$.

Proof: Let $v \in C_1(H_{n+1})$. Since $A_1 \times B_1 \times \ldots \times A_n \times B_n$ is σ-compact, we infer from the proof of the theorem of Baire (cf. Ash [1], p. 390) that there exists an increasing sequence (v_m) of functions $v_m \in M(H_{n+1})$ such that $(a_1, b_1, \ldots, a_n, b_n) \to v_m(h_{n+1})$ is continuous for all m and $v = sup_m v_m$. As in the proof of Lemma 5.2 we conclude that $(a_1, b_1, \ldots, a_{n-1}, b_{n-1}, a, \nu) \to (q_n v)(h_n, a, \nu)$ is lower semi-continuous. Now let A'_n be any denumerable dense subset of A_n. Then we may write

$$(U_n v)(h) = \inf_{\nu} \sup_{a \in A'_n} (q_n v)(h, a, \nu).$$

Theorem 4.2(a) and the compactness of B_n imply that $U_n v$ belongs to $C_1(H_n)$. ◇

Theorem 5.5: If the conditions (S1) are satisfied then $V_n \in C_1(H_n)$ for all n and the basic statements I, II, III and IV are valid.

The *proof* follows along the same lines as the proof of Theorem 5.3 and is therefore not given here.

We remark that for zero-sum Markovian games with separable utility function the basic statements can be proven without the assumption that the spaces A_n are σ-compact.

5.3. Conditions (S2): Denote by $C_2(H_n)$ the set of all bounded functions $v \in M(H_n)$ such that

$$(a_1, a_2, \ldots, a_{n-1}) \to v(h_n) \text{ is upper semi-continuous and}$$
$$(b_1, b_2, \ldots, b_{n-1}) \to v(h_n) \text{ is lower semi-continuous.}$$

Throughout this subsection we impose

Conditions (S2):

(1) A_n and B_n are compact, $n \in \mathbf{N}$.

(2) q_n is s-continuous, $n \in \mathbf{N}$.

(3) $u_n \nearrow u$ for some (u_n) such that $u_n \in C_2(H_n), n \in \mathbf{N}$.

Lemma 5.6: U_n is a map from $C_2(H_{n+1})$ into $C_2(H_n)$.

Proof: Let $v \in C_2(H_{n+1})$. As in the proof of Lemma 5.4 we obtain that $(a_1, \dots, a_{n-1}, \mu) \to (q_n v)(h_n, \mu, \nu)$ is upper semi-continuous and $(b_1, \dots, b_{n-1}, \nu) \to (q_n v)(h_n, \mu, \nu)$ is lower semi-continuous. Then $U_n v \in C_2(H_n)$ by Theorem 4.2(a) and the compactness of A_n and B_n. ◇

Theorem 5.7: Let the conditions (S2) be satisfied.

(a) $V_n \in M(H_n)$ for all n and the basic statements I, II, III and IV are valid.

(b) Let u be bounded. If (u_n) converges uniformly to u then $V_n \in C_2(H_n)$ for all n and there exists a strongly optimal policy for player I.

Proof: (a) By using Theorem 4.3 instead of Theorem 4.2 and appealing to Lemma 5.6 instead of Lemma 5.2, the proof is analogous to the proof of Theorem 5.3. The existence of a measurable map g_n with property (5.5) follows from Theorem 4.3(b) and the inherent m-stage dynamic games ($m \in \mathbf{N}$). For a detailed proof see Rieder [17].

(b) By interchanging the role of the two players we conclude from (a) that there exists a strongly optimal policy for player I. Lemma 4.1(a) tells us that $V_n \in C_2(H_n)$. ◇

References

[1] *Ash, R.B.* (1972). "Real Analysis and Probability". New, York,

[2] *Fan, K.* (1952). Fixed-Point and Minimax Theorems in Locally Convex Topological Linear Spaces. *Proc. Nat. Akad. Sci.*, **38**, 121-126.

[3] *Federgruen, A.* (1978). On N-Person Stochastic Games with a Denumerable State Space. *Adv. Appl. Prob.*, **10**, 452-471.

[4] *Frid, E.B.* (1973). On Stochastic Games. *Theory Prob. Appl.*, **18**, 389-393.

[5] *Glicksburg, I.* (1952). A Further Generalization of the Kakutani Fixed Point Theorem with Application to Nash Equilibrium Points. *Amer. Math. Soc.*, **3**, 170-174.

[6] *Hinderer, K.* (1970). "Foundation of Non-Stationary Dynamic Programming with Discrete Time Parameter". *Lecture Notes in Operations Research and Mathematical Systems*, **33**, Berlin.

[7] *Iwamoto, S.* (1975). Finite Horizon Markov Games with Recursive Payoff Systems. *Memoirs of the Fac. Sci., Kyushu Univ.*, **29**, 123-147.

[8] *Kindler, J.* (1975). Spiele mit abzählbarem Baum. *Operations Research Verfahren*, **21**, 141-154.

[9] *Küenle, H.-U.* (1986). Stochasticsche Spiele und Entscheidungsmodelle. Leipzig.

[10] *Maitra, A. and T. Parthasarathy* (1970). On Stochastic Games. *J. Optim. Theory Appl.*, **5**, 289-300.

[11] *Maitra, A. and T. Parthasarathy* (1971) On Stochastic Games II. *J. Optim. Theory Appl.*, **8**, 154-160.

[12] *Nowak, A.S.* (1984). Approximation Theorems for Zero-Sum Non-Stationary Stochastic Games. *Proc. Amer. Math. Soc.*, **92**, 418-424.

[13] *Nowak, A.S.* (1984). Noncooperative Non-Stationary Stochastic Sames. *OPSEARCH*, **21**, 199-208.

[14] *Nowak, A.S.* (1986). Semicontinuous Non-Stationary Stochastic Games. *J. Math. Anal. Appl.*, **117**, 84-89.

[15] *Parthasarathy, T.* (1973). Discounted, Positive and Noncooperative Stochastic Games. *Intern. J. Game Theory*, **2**, 25-37.

[16] *Parthasarathy, T. and M. Stern* (1977). Markov Games - a Survey. in: E. Roxin, P. Liu and R. Sternberg (eds.) *Differential Games and Control Theory, II*, New York, 1-46.

[17] *Rieder, U.* (1978). On Semi-Continuous Dynamic Games. Technical Report, University of Karlsruhe.

[18] *Rieder, U.* (1979). Equilibrium Plans for Nonzero-Sum Markov Games. in: O. Moeschlin and D. Pallaschke (eds.) *Seminar on Game Theory and Related Topics*, Amsterdam.

[19] *Schäl, M.* (1975). Conditions for Optimality in Dynamic Programming and for the Limit of n-Stage Optimal Policies to Be Optimal. *Zeitschrift Wahrscheinlichkeitstheorie*, **32**, 179-196.

[20] *Schäl, M.* (1981). Stochastic Non-Stationary Two-Person Zero-Sum Games. *Z. Angew. Math. Mech.*, **61**, 352-353.

[21] *Sengupta, S.K.* (1975). Lower Semicontinuous Stochastic Games with Imperfect Information. *Ann. Statist.*, **3**, 554-558.

[22] *Shapley, L.S.* (1953). Stochastic Games. *Proc. Nat. Acad. Sci.*, **39**, 1095-1100.

Ulrich Rieder
Department of Mathematics VII
University of Ulm, D-7900 Ulm
W. Germany

Symmetric Stochastic Games of Resource Extraction: The Existence of Non-Randomized Stationary Equilibrium

Mukul Majumdar and Rangarajan Sundaram*

1. Introduction:

1.1 Stochastic Games: A Description

A two-person discounted stochastic game (see, e.g., [15] and [17] for related references) is described by a tuple $\{S, A_1(s), A_2(s), q, r_1, r_2, \beta\}$ having the following interpretation: S, a non-empty Borel subset of a Polish space, is the set of all states of the system; $A_i(s)$, a non-empty Borel subset of a Polish space, is the set of actions available to player $i(= 1, 2)$, when the state is $s \in S$. It is typically assumed that for each $i = 1, 2$, $A_i(s) \subset A_i$ for all $s \in S$, where the A_i's are themselves Borel subsets of Polish spaces. q defines the law of motion of the system by associating (Borel-measurably) with each triple $(s, a_1, a_2) \in S \times A_1 \times A_2$ a probability measure $q(\cdot|s, a_1, a_2)$ on the Borel subsets of S. r_1 and r_2 are bounded measurable functions on $S \times A_1 \times A_2$; the function r_i is the instantaneous reward function for player i. Lastly, β is the discount factor the players employ. Periodically, the players observe a state $s \in S$ and pick actions $a_i \in A_i(s)$, $i = 1, 2$;

---

*We are indebted to Professor Debraj Ray for drawing our attention to the methods of proof employed in showing existence of equilibrium in the related problem of altruistic growth models (see, e.g., [1, 18] and the references there), and for several helpful conversations. Thanks are also due to Professors Prajit Dutta and T. Parthasarathy for going over the proof with us. Research support from the National Science Foundation is gratefully acknowledged. Detailed proofs of some results omitted in the interest of brevity may be found in the Working Paper [11] circulated by the CAE, Cornell University. The usual disclaimer applies.

this choice of actions is made with full knowledge of the game's history. As a consequence of the chosen actions, two things happen: firstly, the players receive rewards of $r_1(s, a_1, a_2)$ and $r_2(s, a_1, a_2)$ respectively. Secondly, the system moves to a new state s' according to the distribution $q(\cdot|s, a_1, a_2)$. The process is then repeated from the state s', and so on *ad infinitum*. The objective of each player is to maximize expected payoffs over the infinite duration of the game.

Let $h_t = \{s_0, a_{10}, a_{20}, \ldots, s_{t-1}, a_{1,t-1}, a_{2,t-1}, s_t\}$ denote a generic *history* of the game up to period t, and let H_t denote the set of all possible histories up to t. Let $P(A_i(s))$ and $P(A_i)$ be the set of all probability distributions on $A_i(s)$ and A_i respectively, $i = 1, 2$. A *strategy* $\sum_i$ for player i is a sequence of functions $\{\sigma_{it}\}$, where for each t, σ_{it} specifies an action for player i by associating (Borel measurably) with each history h_t, an element of $P(A_i(s_t))$. A strategy $\sum_i$ for player i is (non-randomized) *stationary* if there is a Borel function $\sigma_i : S \to A_i$ such that $\sigma_i(s) \in A_i(s)$ for all $s \in S$, and $\sigma_{it}(h_t) = \sigma_i(s_t)$ for all h_t and for all t. We shall refer to the function σ_i as a *policy function*, and when talking about a non-randomized stationary strategy, we also refer to it by the associated policy function.

A pair $(\sum_1, \sum_2)$ of strategies for players 1 and 2 respectively, associates with each initial state s, a tth-period expected reward $r_{it}(\sum_1, \sum_2)(s)$ for player i determined by the functions r_1 and r_2. The total expected reward for player i, denoted $I_i(\sum_1, \sum_2)(s)$ is then

$$I_i(\Sigma_1, \Sigma_2)(s) = \sum_{t=0}^{\infty} \beta^t r_{it}(\Sigma_1, \Sigma_2)(s).$$

A strategy $\sum_1^*$ is *optimal* for player 1 (or, constitutes a *best-response* (BR) to $\sum_2$) if $I_1(\sum_1^*, \sum_2)(s) \geq I_1(\sum_1, \sum_2)(s)$ for all $\sum_1$ and s. Similarly, a BR to $\sum_1$ is defined for player 2. A *Nash equilibrium* (or, simply, equilibrium) to the stochastic game is a pair of strategies $(\sum_1^*, \sum_2^*)$ such that for $i = 1, 2$, $\sum_i^*$ is a BR to $\sum_j^*$, $j \neq i$.

1.2. Summary of the main results:

This paper considers a special class of stochastic games allowing for a continuum of states *and* actions. The sets of states and actions are required to satisfy certain restrictions, as is the stochastic process that determines the law of motion q. The special structure is motivated by models in the economic theory of non-cooperative extraction of common-property resources.[1] While a brief explanation of this link is provided in subsection 2.2, a detailed explanation (in the context of a deterministic game) may be found in [24].

[1]This problem has been studied in a deterministic framework quite extensively, but by using specific functional forms - see e.g., [9].

The imposition of a certain symmetry in the payoff functions (equation ($R1$) below) in addition to the restrictions mentioned above enables us to prove the following strong results: there is an equilibrium in (non-randomized) *stationary* strategies to the class of games considered in the paper. Further, the policy functions associated with the equilibrium can be chosen to be *lower-semicontinuous* functions, with slopes bounded above by 1/2.[2] The sharpness of this result is to be contrasted with the available results in the literature, where existence is typically shown in randomized strategies that cannot be easily characterized. The price paid for obtaining this result is that the model is more restrictive than the standard models in, for example, [8,15,16].

2. The Model:

2.1. Notation and Definitions

The set of all real numbers (resp. non-negative reals, strictly positive reals) is denoted by $\mathcal{R}$ (resp. $\mathcal{R}_+, \mathcal{R}_{++}$). The n-fold Cartesian product of $\mathcal{R}, \mathcal{R}_+$, and $\mathcal{R}_{++}$ are denoted by $\mathcal{R}^n, \mathcal{R}^n_+$, and $\mathcal{R}^n_{++}$, respectively. For any set $X, 2^X$ denotes the set of all non-empty subsets of X.

Let $S = A_1 = A_2 = [0,1]$. Define the feasible action correspondences for player $i = 1,2$, by $A_i(s) = [0,s]$. Clearly, the feasible action correspondences are continuous on S.[3]

Before proceeding to describe the formal structure of the game, we present an informal interpretation of its components. The non-negative number s denotes the available stock of a common-property resource, while a_i represents player i's planned extraction of the resource. (Both players are assumed to know s and the other player's plan.) If plans are feasible (i.e., if $a_1 + a_2 \leq s$) then they are carried out and player i receives a reward ("utility" in intertemporal-economics parlance) of $u(a_i)$. If plans are infeasible ($a_1 + a_2 > s$) then we assume ad hoc that each player extracts half the available stock of the resource and receives a reward of $u(s/2)$. We shall have more to say about this ad hoc assumption shortly.

Given (s, a_1, a_2) the function $h(s, a_1, a_2) = max\{0, s - a_1 - a_2\}$ determines the 'investment' level, the amount left over after extraction by the players. This investment is transformed stochastically into next-period's available stock s', for example, through a 'renewal' function f, and the realization of a random variable r, as $s' = f(h(s, a_1, a_2), r)$. The functions f and h, combined with the distribution

---

[2]A real-valued function f is lower-semicontinuous or lsc [resp. upper-semicontinuous, or usc] at a point x in its domain if for all $x_n \to x$, it is the case that liminf $f(x_n) \geq f(x)$ [resp. lim sup $f(x_n) \leq f(x)$]. Note that f is continuous at x iff it is both usc and lsc at x.

[3]For the definition of a continuous correspondence, (as well as for those of upper-semicontinuous (usc) and lower-semicontinuous (lsc) correspondences), the reader is referred to [6].

of r yields a (conditional) probability distribution of s' given (s, a_1, a_2). We denote this conditional distribution by q and, rather than imposing assumptions on f and r, impose restrictions directly on q.

Departing from standard practice we define the transition mechanism q as a (conditional) probability distribution function on $\mathcal{R}_+$, given $(s, a_1, a_2) \in \mathcal{R}^3_+$, so that if s^* denotes next period's realization given (s, a_1, a_2), then $q(s'|s, a_1, a_2) = Pr\{s^*|s, a_1, a_2\}$. It will follow from the restrictions we place on q that if $s \in S$, $a_i \in A_i(s)$, then $q(1|s, a_1, a_2) = 1$, so next period's stock is also in S w.p.1.

For simplicity denote the vector (s, a_1, a_2) by $y \in \mathcal{R}^3_+$ and $h(s, a_1, a_2)$ by $h(y)$. The restrictions on q are (i) a 'boundedness' condition that with each investment level today is associated an upper bound on the stock available tomorrow, (ii) strictly positive investments today yield strictly positive stocks tomorrow and (iii) no free production.

(Q1) (i) For each $y \in \mathcal{R}^3_+$, there is $s(y) \in \mathcal{R}_+$ such that $q(s(y)|y) = 1$.

(ii) If $h(y) > 0$, then $\inf\{s' : q(s'|y) > 0\} \in \mathcal{R}_{++}$. Further, in this case, $q(\cdot|y)$ is continuous on $\mathcal{R}$.

(iii) If $h(y) = 0$, *then* $q(0|y) = 1$.

We also assume that higher investments yield probabilistically higher stock levels.

(Q2) If $h(\overline{y}) > h(y)$, then $q(s'|\overline{y}) \leq q(s'|y)$ for all $s' \in \mathcal{R}_+$

The next two assumptions are concerned with reproductivity of the resource. Assumption (Q3) requires the existence of a maximum sustainable stock (set equal to unity by a suitable choice of measurement-units), while (Q4) implies that for a positive but sufficiently small level of investment, with probability one, the stock tomorrow is no less than the investment today (usually referred to as a "productivity" or Inada condition). Formally:

(Q3) If $h(y) \geq 1$, then $q(h(y)|y) = 1$.

(Q4) There is $\eta \in (0, 1)$ such that if $0 < h(y) < \eta$, then $q(h(y)|y) = 0$.

Finally, the standard weak continuity of the law of motion q:

(Q5) If $y^n \to y$, then the sequence of distribution functions $q(\cdot|y^n)$ converges weakly to the distribution function $q(\cdot|y)$.

Example: Let λ be uniformly distributed on [1,2], and let $f(x) = \frac{1}{2}\sqrt{x}, x \geq 0$. Define $q(s'|y) = Pr\{\lambda f(h(y)) \leq s'\}$. If $h(y) \geq 1$, $\lambda f(h(y)) = \frac{\lambda}{2}\sqrt{h(y)} \leq h(y)$, so $q(h(y)|y) = 1$. If $h(y) < \eta = \frac{1}{4}$, then $q(h(y)|y) = Pr\{\lambda f(h(y)) \leq h(y)\} = Pr\{\frac{\lambda}{2}\sqrt{h(y)} \leq h(y)\} = Pr\{\lambda \leq 2\sqrt{h(y)}\} = 0$. Similarly, the other conditions are verified.

Next, let $u : \mathcal{R}_+ \to \mathcal{R}_+$ be a function satisfying the following condition:

(U1) u is strictly concave, strictly increasing and continuous on $\mathcal{R}_+$; u is continuously differentiable on $\mathcal{R}_{++}$ and satisfies $lim_{c\downarrow 0}\, u'(c) = \infty$.

Example: $u(c) = c^{\alpha}$, $\alpha \in (0,1)$.

The reward functions, like the transition mechanism, are defined for all $(s, a_1, a_2) \in \mathcal{R}^3_+$, and are given by:

$$\text{(R1)} \quad \begin{aligned} r_i(s, a_1, a_2) &= u(a_i), \text{ if } s - a_1 - a_2 \geq 0 \\ &= u(s/2), \text{ otherwise} \end{aligned}.$$

Assumption (R1) forms the basic symmetry assumption on the game, crucial in showing the existence of equilibrium. In what follows, (Q1) through (Q5), (U1) and (R1) are always assumed.

2.2 The Main Theorem

Note that by assumption (Q3), for any $s \in S$, and any $a_i \in A_i(s)$, it is the case that $s' \in S$ w.p.1. Further if $s \in [0, \overline{y}]$ for any $\overline{y} > 1$, and if $0 \leq a_1, a_2 \leq s$, then $s' \in [0, \overline{y}]$ w.p.1. The first of these observations implies that the game is well-defined. The second observation is important to show existence of equilibrium as we shall explain in section 4.

Furthermore, owing to the *ad hoc* nature of the infeasibility rule, trivial equilibria always exist, i.e., equilibria in which players plan to extract more than the available stock in some period after some history. [4] Indeed, it is easily checked that the non-randomized strategies (γ_1, γ_2) defined by $\gamma_1(s) = s = \gamma_2(s)$ in each period constitutes an equilibrium to the game. The main result in this paper is the demonstration of existence of a *non-trivial* equilibrium in non-randomized stationary strategies (γ_1^*, γ_2^*), i.e., an equilibrium in which at each $s > 0$, it is the case that $(\gamma_1^*(s) + \gamma_2^*(s)) < s$. This equilibrium is independent of the infeasibility rule employed, as will be demonstrated in section 4. Formally, we can state the main:

Existence Theorem: Under (Q1) through (Q5), (U1) and (R1), the stochastic game has an equilibrium (γ_1^*, γ_2^*) in non-randomized stationary strategies satisfying:

(i) $0 < \gamma_1^*(s) + \gamma_2^*(s) < s$ *for all* $s > 0$;

(ii) (γ_1^*, γ_2^*) are lower semicontinuous on S;

(iii) *for* i = 1,2 and for all s, s' *in* S, $s \neq s'$,

$$\frac{\gamma_i^*(s) - \gamma_i^*(s')}{s - s'} \leq \frac{1}{2}.$$

The next two sections contain an outline of the proof. Some informal remarks on the strategy we adopt may be useful: first, the game is transformed to a "generalized game" in the sense of [5] by making action spaces dependent. This makes the outcome in the generalized game independent of the ad hoc infeasibility rule. Most of

[4] We refer to such equilibria as "trivial" since they depend in an essential way on the infeasibility rule employed. Our desire to find nontrivial equilibria is also motivated in part by the question: under what conditions is it the case that resources are not driven to extinction in finite time? For more on this and related questions, see [4] and the references cited therein.

sections 3 and 4 is concerned with establishing an equilibrium to the generalized game. That the equilibrium satisfies condition (i) [hence, is independent of the ad hoc infeasibility rule, and that the resource is not extinct in finite time] is shown in Lemma 4.9 and leads to the conclusion that the equilibrium of the generalized game is also the equilibrium of the original game.

3. The Best-Response Map: For simplicity of notation we omit player subscripts from what follows. Let $\gamma : S \to S$ be a measurable function satisfying $\gamma(s) \in [0, s]$ for each $s \in S$. Each such γ defines a non-randomized stationary strategy for player i. Given γ define player j's $(j \neq i)$ feasible action correspondence by $A_j(\gamma)(s) = [0, s - \gamma(s)]$. Now note that in maximizing total expected payoff, player j faces a stationary environment: the functions u, q, and γ are invariant with time (q is now simply $q(\cdot|s, \gamma(s)$, a) from the point of view of player j). Thus in seeking the optimal solution to such a problem, by Lemma 2 in [3] player j can restrict attention to non-randomized strategies. Let $G(\gamma)$ represent the set of all such strategies. Each strategy $\tilde{\gamma} \in G(\gamma)$ must satisfy (by the dependence of the action spaces), the condition $\tilde{\gamma}(h_t) \leq [s_t - \gamma(s_t)]$ where $h_t = (s_0, a_{10}, a_{20}, \ldots, s_{t-1}, a_{1,t-1}, a_{2,t-1}, s_t)$ is the history of the game up to period t, and the actions $(a_{1\tau}, a_{2\tau})$ for $\tau \leq t$ are determined by γ and $\tilde{\gamma}$. Each strategy $\tilde{\gamma} \in G(\gamma)$ also yields an expected payoff to player j that we shall denote by $W_\gamma(\tilde{\gamma})(s)$, where $s_0 = s$ (the subscript γ of W denotes the dependence of player j's actions - hence his expected payoffs - on γ). A strategy $\tilde{\gamma}^* \in G(\gamma)$ is *optimal* and $\tilde{\gamma}^*$ constitutes a *generalized best-response* (GBR) to γ if $W_\gamma(\tilde{\gamma}^*)(s) \geq W_\gamma(\tilde{\gamma})(s)$, for all $s \in S$, for all $\tilde{\gamma} \in G(\gamma)$. That is, a GBR is a strategy $\tilde{\gamma}^* \in G(\gamma)$ that solves for all $s \in S$

(P) $max_{\{\tilde{\gamma} \in G(\gamma)\}} W_\gamma(\tilde{\gamma})(s)$, given q, γ.

If such a $\tilde{\gamma}^*$ exists (of course, it need not always), then $W_\gamma(\tilde{\gamma}^*)$ is referred to as player j's *value function* from optimally responding to γ. Conditions to ensure that a GBR exists are presented below.

Theorem 3.1: Suppose $\gamma : S \to S$ is a lower-semicontinuous (lsc) function on S satisfying $\gamma(s) \in [0, s]$ for each $s \in S$ and further, for all $s_1 \neq s_2$, $[(\gamma(s_1) - \gamma(s_2))/(s_1 - s_2)] \leq 1$. Then, problem (P) is well-defined: there is a Borel function $\hat{\gamma}^* : S \to S$ such that $\hat{\gamma}^*$ is optimal in $G(\gamma)$, i.e., player i has a stationary GBR to γ. Furthermore, the value function $W_\gamma(\hat{\gamma}^*)$ (henceforth denoted by V_γ) is upper-semicontinuous (usc) on S.

A detailed proof of this result may be found in [11].

4. The Existence of Equilibrium: It follows from Theorem 3.1 that *if* we could show that lsc policy functions γ possessed lsc GBR functions $\hat{\gamma}$, an equilibrium to the generalized game could be obtained by using a standard Debreu-Nash fixed-point argument on the space of lsc functions (endowed with a suitable topology). Unfortunately, it is easy to show the existence of lsc functions that do *not* possess lsc GBR functions.[5]

We employ therefore a completely different approach, one in which the symmetry in the payoff functions is exploited to provide the equilibrium. As the first step in the process, we expand S to a larger space $\overline{S} = [0, \overline{\overline{y}}]$ for $\overline{\overline{y}} > 1$. The equilibrium is constructed on $[0, \overline{\overline{y}}]$, and it is shown below (see Lemma (4.8)) that the restriction of the equilibrium strategies to S is an equilibrium on S. Note that by (Q1) - (Q5), if $s \in \overline{S}$, then $s' \in \overline{S}$ w.p.1. Consider the following space of functions on $\overline{S}$:

$$\Psi = \{\psi : \overline{S} \to \overline{S} \mid \psi \text{ is } usc \text{ and non-decreasing on } \overline{S},$$
$$\psi(\overline{\overline{y}}) = \overline{\overline{y}}, \text{ and } \psi(s) \in [0, s] \text{ for all } s \in \overline{S}\}.$$

Each $\psi \in \Psi$ defines a (non-randomized stationary) strategy $\gamma(\psi)$ for player 1 by the rule:

$$\gamma(\psi)(s) = \frac{1}{2}(s - \psi(s)).$$

Since ψ is usc, non-decreasing, so $\gamma(\psi)$ satisfies the conditions of Theorem 3.1 (which of course is not affected by expanding the state and action spaces to $\overline{S}$ from S), and there exists a GBR denoted by $\hat{\gamma}(\psi)$. Define $\hat{\psi} : \overline{S} \to \overline{S}$ by

$$\hat{\psi}(s) = s - \gamma(\psi)(s) - \hat{\gamma}(\psi)(s).$$

In Lemma 4.3 below, it is shown that there exists a $\hat{\gamma}(\psi)$, a unique GBR to each $\gamma(\psi)$ such that $\hat{\psi}$ defined thus is in Ψ. This defines a map from Ψ into itself. Consider a fixed-point of this map. At such a point, $\hat{\psi} = \psi$, so from the above equations, some manipulation yields $\hat{\gamma}(\psi) = \gamma(\psi)$ or $\gamma(\psi)$ *is a GBR to itself on* $\overline{S}$. Lemmas 4.8-4.10 then conclude the proof by showing that it is in fact the case that $\gamma(\psi)$is a *best-response* to itself when the state space is restricted to S. By the symmetry of the payoffs (equation R1) the argument is complete.

These ideas underlie the following results but rather than invoking the functions $\gamma(\psi)$ and $\hat{\gamma}(\psi)$, notation is simplified as follows: player 2's actions in response to $\gamma(\psi)$ are now interpreted as the investment level he chooses given player 1's action, so that if he takes an action $a > 0$, his instantaneous reward is given by $u(s - \gamma(\psi)(s) - a)$. Define $R_\psi(s) = \frac{1}{2}(s + \psi(s))$ for $s \in S$, $\psi \in \Psi$. Note that the conditional distribution over $\overline{S}$ of next period's state s' depends now only on a. Abusing notation we denote this distribution by $q(\cdot|a)$. Finally, let V_ψ denote

[5]A trivial example is the following: let $\gamma(s) = s$ for $s \in [0, 1)$ and $\gamma(1) = 0$. The unique GBR is $\gamma(s) = 0$ for $s \in [0, 1)$ and $\gamma(1) = 1$, which is not lsc at $s = 1$.

player 2's value function from a GBR to $\gamma(\psi)$. We rewrite the Bellman Optimality equation in this notation as:

$$V_\psi(s) = \max_{a\in[0,R_\psi(s)]}\{u(R_\psi(s) - a) + \beta \int V_\psi(s')dq(s'|a)\} \tag{4.1}$$

Let $\widehat{V}$ denote the function V_ψ when $\psi(s) \equiv s$ for all $s \in \overline{S}$. Then, clearly, for any $\psi \in \Psi$, $V_\psi \leq \widehat{V}$. Define

$$\Omega = \{v : \overline{S} \to R_+ \ |v \text{ is } usc \text{ and non-decreasing on } \overline{S},$$
$$v(0) = \frac{u(0)}{1-\beta},\ v(\overline{\overline{y}}) = \frac{u(\overline{\overline{y}})}{1-\beta},\ v \leq \widehat{V}\}$$

Endow Ψ, Ω with the topology of weak-convergence (i.e., pointwise convergence to continuity points of the limit function - see e.g., [2]). We can then show

Lemma 4.1: Ψ and Ω are convex compact metric spaces. Further, Ψ has the fixed-point property.

Proof: Convexity is obvious. To see compactness of Ψ consider the set N of finite measures ν on the Borel sets of $\overline{S}$ satisfying $\nu(\overline{S}) = \overline{\overline{y}}$ for all $\nu \in N$. Since $\overline{S}$ is compact metric, a well known result establishes that N endowed with the topology of weak convergence (weak topology, for short) is also a compact metric space (see, e.g., [14]). If Ψ_0 denotes the set of distribution functions corresponding to measures in N, it follows that Ψ_0 is also a compact metric space under the weak topology. Since Ψ is a closed subset of Ψ_0 it also has this property. That it possesses the fixed-point property follows from the Schauder-Tychonoff theorem (see, e.g., [21]), whose conditions are easily seen to be met.

Ω is similarly a compact metric space if we can show it to be closed in the weak topology. Since $\widehat{V}$ corresponds to the value function of a one-person dynamic programming problem with (weakly)-continuous transition and continuous payoffs, it is straightforward to show that $\widehat{V}$ is itself a continuous function. By the assumptions on q, $\widehat{V}(0) = u(0)/(1-\beta)$. Since $v \leq \widehat{V}$ for all $v \in \Omega$, the result readily follows. $\diamond$

Now observe that for fixed ψ, the feasible action correspondence $[0, R_\psi(s)]$ is increasing in s, i.e., any action feasible at s_1 is also feasible at s_2 if $s_2 > s_1$. Since u is increasing in its argument, it is immediate by the upper-semicontinuity of V_ψ that

Lemma 4.2: For each ψ, V_ψ is non-decreasing and right-continuous on $\overline{S}$.

Now for each ψ redefine the value of V_ψ at $\overline{\overline{y}}$ by setting $V_\psi(\overline{\overline{y}}) = u(\overline{\overline{y}})/(1-\beta)$. Thus defined, V_ψ still satisfies the conditions of Lemma 4.2, *therefore* $V_\psi \in \Omega$ *for each* $\psi \in \Psi$.

As the second step in the proof we shall now construct a map from Ψ into itself. To this end, we define for $\psi \in \Psi$ and $v \in \Omega$ a map $F_{\psi,v} : \overline{S} \to 2^{\overline{S}}$ by $F_{\psi,v}(\overline{\overline{y}}) = \overline{\overline{y}}$, and for $0 \leq s < \overline{\overline{y}}$,

$$F_{\psi,v}(s) = \text{argmax}_{a \in [0, R_\psi(s)]}\{u(R_\psi(s) - a) + \beta \int v(s')dq(s'|a)\}.$$

If $v = V_\psi$, then we shall write F_ψ for $F_{\psi,v}$.

By Lemma 2.1 and Theorem 2.1 in [15], $F_{\psi,v}$ is well-defined and a measurable correspondence, and further admits a measurable selection. (This is a consequence of the fact that $v \in \Omega$ implies $\int v(s')dq(s'|a)$ is *usc* as a function of a; see [11]). In fact, we can show that

Lemma 4.3: There is a unique selection $\widehat{\psi}$ from $F_{\psi,v}$ such that $\widehat{\psi} \in \Psi$.

Proof: The lemma is proved in 3 steps:

Claim 1: If $s_1 > s_2$, $a_1 \in F_{\psi,v}(s_1)$, $a_2 \in F_{\psi,v}(s_2)$, then $a_1 \geq a_2$.

This is proved by a standard argument in intertemporal economics that relies upon the strict concavity of u. See [11] for details.

Claim 2: If $s_n \downarrow s$, $a_n \in F_{\psi,v}(s_n)$, $a_n \to a$, then $a \in F_{\psi,v}(s)$.

Proof: Suppose, contrary to the claim, it was the case that $a \notin F_{\psi,v}(s)$. Since the latter is non-empty it contains $\widehat{a} \leq s - \gamma(\psi)(s)$ such that

$$\begin{aligned} u(R_\psi(s) - \widehat{a}) + \beta \int v(s')dq(s'|\widehat{a}) \\ > u(R_\psi(s) - a) + \beta \int v(s')dq(s'|a). \end{aligned} \tag{4.2}$$

Since $a_n \downarrow a$, assumption (Q5) implies the weak convergence of $q(\cdot|a_n)$ to $q(\cdot|a)$. Since v is *usc*, $\limsup_{n\to\infty} \int v(s')dq(s'|a_n) \leq \int v(s')dq(s'|a)$, so combining this with equation (4.2) and the fact that right-continuity of ψ and continuity of u together imply $u(R_\psi(s_n) - a_n) \to u(R_\psi(s) - a)$, we obtain the existence of $\alpha > 0$ such that for large n

$$\begin{aligned} u(R_\psi(s_n) - a_n) + \beta \int v(s')dq(s'|a_n) + 2\alpha \\ < u(R_\psi(s) - \widehat{a}) + \beta \int v(s')dq(s'|\widehat{a}.) \end{aligned} \tag{4.3}$$

Using the additional fact that $u(R_\psi(s_n) - \hat{a}) \to u(R_\psi(s) - \hat{a})$, (4.3) in turn implies that for all sufficiently large n

$$u(R_\psi(s_n) - a_n) + \beta \int v(s')dq(s'|a_n) + \alpha \tag{4.4}$$
$$< u(R_\psi(s_n) - \hat{a}) + \beta \int v(s')dq(s'|\hat{a}).$$

But $\hat{a} \leq s - \gamma(\psi)(s) \leq s_n - \gamma(\psi)(s_n)$, so $\hat{a}$ is feasible at s_n. Equation (4.4) therefore contradicts the optimality of a_n for all large n.

Note that by Claim 2, $max\{F_{\psi,v}(s)\}$ is well-defined at each $s \in [0, \overline{\overline{y}}]$. Defining $\widehat{\psi}(s) = max\{F_{\psi,v}(s)\}$ for $s \in S$, we see that Claims 1 and 2 together imply that $\widehat{\psi}$ is right-continuous and non-decreasing. Therefore $\widehat{\psi}$ is usc on S, and $\widehat{\psi} \in \Psi$, since $F_{\psi,v}(\overline{\overline{y}})) = \overline{\overline{y}}$. The last step in the proof of Lemma 4.3 is

Claim 3: $\widehat{\psi}$ is the only usc selection from $F_{\psi,v}$.

Proof: Suppose there were another usc selection $\overline{\psi}$. Note that $\overline{\psi}$ is non-decreasing, hence right-continuous. Since $\widehat{\psi} \neq \overline{\psi}$, there is $s \in \overline{S}$ such that $\widehat{\psi}(s) \neq \overline{\psi}(s)$, so $\widehat{\psi}(s) > \overline{\psi}(s)$. Let $s_n \downarrow s$. Then $\overline{\psi}(s_n) \downarrow \overline{\psi}(s)$, so for large enough n, $\widehat{\psi}(s) > \overline{\psi}(s_n) \in F_{\psi,v}(s_n)$, but $s < s_n$ and $\widehat{\psi}(s) \in F_{\psi,v}(s)$, so this contradicts Claim 1. ◇

Note that if $v = V_\psi$, then Lemma 4.3 implies that for each $\psi \in \Psi$, there is a $GBR\hat{\gamma}(\psi)$ to $\gamma(\psi)$ such that the resulting 'savings' function $\widehat{\psi}(s) = s - \gamma(\psi)(s) - \hat{\gamma}(\psi)(s)$ is in Ψ. Thus, Lemma 4.3 defines a map from Ψ into itself. A fixed-point ψ^* of this map yields a pair of functions $(\gamma*, \gamma*)$ defined by $\gamma^*(s) = 1/2(s - \psi^*(s))$ such that γ^* is a GBR to itself on $[0, \overline{\overline{y}})$. Since Ψ possesses the fixed-point property (Lemma 4.1), the continuity of the map $B : \Psi \to \Psi$, $B(\psi)(s) = \widehat{\psi}(s) = max\{F_\psi(s)\}$ will provide us with the desired fixed-point. A few preliminary results are needed first:

Lemma 4.4: Let $\widehat{\psi}$ be the unique selection from $F_{\psi,v}$ satisfying $\widehat{\psi} \in \Psi$. If $\widehat{\psi}$ is continuous at $s \in [0, \overline{\overline{y}})$, then $F_{\psi,v}$ is single-valued at s.

Proof: Suppose not. Let $\widehat{\psi}(s) > a \in F_{\psi,v}(s)$. Let $s_n < s$, $s_n \to s$. By continuity of $\widehat{\psi}$ at s, $\widehat{\psi}(s_n) \to \widehat{\psi}(s)$. So for large n, $\widehat{\psi}(s_n) > a$, but $a \in F_{\psi,v}(s)$, and $s > s_n$, a contradiction to Claim 1 in Lemma 4.3. ◇

Lemma 4.5: Suppose $\psi_n \to \psi \in \Psi$, $s_n \to s \in \overline{S}$. Then,

(i) $lim\,sup\,\psi_n(s_n) \leq \psi(s)$

(ii) if ψ is continuous at s, then $lim_{n\to\infty}\,\psi_n(s_n) = \psi(s)$.

Proof: Follows from a modification of Lemma 5.3 in [1]. See [11] for details.

Lemma 4.6: Suppose $v_n \to v \in \Omega$ and $\psi_n \to \psi \in \Psi$. Suppose also that $s \in \overline{S}$ is a continuity point of ψ. Then,

$$\int v_n(s')dq(s'|\psi_n(s)) \to \int v(s')dq(s'|\psi(s)).$$

Proof: By the generalized Dominated convergence theorem, it suffices to show that (i) $q(\cdot|\psi_n(s))$ converges weakly to $q(\cdot|\psi(s))$, (ii) $\{v_n\}$ is a uniformly integrable sequence, and (iii) $v_n \to v$ in distribution. Since, by hypothesis, s is a continuity point of ψ, so $\psi_n(s) \to \psi(s)$, and (i) follows from assumption (Q5). Since $v_n(s') \leq (1-\beta)^{-1}u(\overline{y})$ for all $s' \in S$, (ii) is immediate. Let μ_n be the measure on $\overline{S}$ corresponding to $q(\cdot|\psi_n(s))$, and μ that corresponding to $q(\cdot|\psi(s))$. Then, we need to show that $\mu_n v_n^{-1}$ converges weakly to μv^{-1}. Since μ_n converges weakly to μ, it suffices by ([2], Theorem 5.5) to show that $\mu(E') = 0$ where $E' = \{s' \in \overline{S}|$ there is $s'_n \to s'$ such that $v_n(s'_n)$ does not converge to $v(s')\}$. Let $E = \{s' \in \overline{S}|v$ is discontinuous at $s'\}$. Lemma 4.5 depends only on the fact that ψ_n and ψ are monotone right-continuous functions. Since functions in Ω also possess this property, clearly $E' \subset E$. Further, E' is measurable by [2, p.226]. Note that $0 \notin E'$, since $s'_n \to 0$ implies by Lemma 4.5 that $lim\,sup_{n\to\infty}\, v_n(s'_n) \leq v(0) = (1-\beta)^{-1}u(0)$, while since $v_n \in \Omega$, $v_n(s'_n) \geq v_n(0) = (1-\beta)^{-1}u(0)$, so $liminf_{n\to\infty}\, v_n(s'_n) \geq (1-\beta)^{-1}u(0) = v(0)$. We identify two cases: (i) $\psi(s) = 0$, so $q(s'|\psi(s)) = 1$ for all $s' \geq 0$. Since $0 \notin E'$, clearly $\mu(E') \leq \mu(E) = 0$ since in this case $\mu(A) = 0$ if $0 \notin A$ for any Borel set A. (ii) $\psi(s) > 0$. By Q1(ii), $q(\cdot|\psi(s))$ is continuous, and its induced measure μ contains no atoms, so (since E is countable by the fact that v is non-decreasing), $\mu(E') \leq \mu(E) = 0$ in this case also. ◇

Lemma 4.7: Suppose k_1, k_2 are non-decreasing, right-continuous functions on $\widehat{S} = [0, \overline{y})$. Suppose also that $\widehat{D}$ is dense in $\widehat{S}$ and $k_1 = k_2$ on $\widehat{D}$. Then, $k_1 = k_2$ on $\widehat{S}$.

Proof: Straightforward.

We are now ready for

Lemma 4.8: $B : \Psi \to \Psi$ is a continuous map when Ψ is endowed with the weak topology.

Proof: Recall that sequential arguments suffice. Let ψ_n be a sequence in Ψ converging (weakly) to $\psi \in \Psi$. Let $\widehat{\psi}_n = B(\psi_n)$ and for notational simplicity denote V_{ψ_n} by V_n. Since Ψ, Ω, are compact, metric, we may assume without loss of generality that $\widehat{\psi}_n \to \widehat{\psi} \in \Psi$, $V_n \to V \in \Omega$. We are required to show that $\widehat{\psi} = B(\psi)$.

As a first step, consider

$$F(s) = \begin{cases} \operatorname{argmax}_{a \in [0, R_\psi(s)]} \{u(R_\psi(s) - a) + \beta \int V(s')dq(s'|a)\} & s \in [0, \overline{\overline{y}}) \\ \overline{\overline{y}} & \text{if } s = \overline{\overline{y}}. \end{cases}$$

By Lemma 4.3, there is a unique $\overline{\psi} \in \Psi$ such that $\overline{\psi}(s) \in F(s)$ for $s \in S$. We claim that $\overline{\psi} = \widehat{\psi}$. Note that to prove this claim, it suffices by Lemma 4.7 to show that $\overline{\psi} = \widehat{\psi}$ on a set dense in $\overline{S}$.

Let D' be the set of discontinuity points of any of the following functions: $\psi_n, \widehat{\psi}_n, \overline{\psi}, \psi, \widehat{\psi}, V_n$, and V. Since each of these functions is monotone (and right-continuous), D' is at most countable. Hence, $D = \overline{S} - D'$ is dense in $\overline{S}$.

We shall show that $\overline{\psi} = \widehat{\psi}$ on D. Let $s \in D$. Consider first the case $\overline{\psi}(s) < R_\psi(s)$. Since ψ is continuous at s, $\psi_n(s) \to \psi(s)$, so $R_{\psi_n}(s) \to R_\psi(s)$, and therefore, for large n, $R_{\psi_n}(s) > \overline{\psi}(s)$. For all such n,

$$\begin{aligned} &u(R_{\psi_n}(s) - \widehat{\psi}_n(s)) + \beta \int V_n(s')dq(s'|\widehat{\psi}_n(s)) \\ &\geq u(R_{\psi_n}(s) - \overline{\psi}(s)) + \beta \int V_n(s')dq(s'|\overline{\psi}(s)). \end{aligned} \tag{4.5}$$

By Lemma 4.6, and since $s \in D$, $\int V_n(s')dq(s'|\widehat{\psi}_n(s)) \to V(s')dq(s'|\widehat{\psi}(s))$, and

$$\int V_n(s')dq(s'|\overline{\psi}(s)) \to \int V(s')dq(s'|\overline{\psi}(s)),$$

so taking limits in (4.5) yields

$$\begin{aligned} &u(R_\psi(s) - \widehat{\psi}(s)) + \beta \int V(s')dq(s'|\widehat{\psi}(s)) \\ &\geq u(R_\psi(s) - \overline{\psi}(s)) + \beta \int V(s')dq(s'|\overline{\psi}(s)). \end{aligned} \tag{4.6}$$

Now suppose $\overline{\psi}(s) = R_\psi(s)$. Then, since

$$\begin{aligned} &u(R_{\psi_n}(s) - \widehat{\psi}(s)) + \beta \int V(s')dq(s'|\widehat{\psi}(s)) \\ &\geq u(0) + \beta \int V_n(s')dq(s'|R_\psi(s)), \end{aligned} \tag{4.7}$$

the same arguments imply that taking limits in (4.7) we obtain

$$\begin{aligned} u(R_\psi(s) - \widehat{\psi}(s)) + \beta \int V(s')dq(s'|\widehat{\psi}(s)) & \\ \geq u(0) + \beta \int V(s')dq(s'|R_\psi(s)) & \\ = u(R_\psi(s) - \overline{\psi}(s)) + \beta \int V(s')dq(s'|\overline{\psi}(s)). & \end{aligned} \tag{4.8}$$

Equations (4.6) and (4.8) imply that $\widehat{\psi}(s) \in F(s)$ for $s \in D$, if $\overline{\psi}(s) \in F(s)$ for $s \in D$. This implies that $\widehat{\psi} = \overline{\psi}$ on D, and by Lemma 4.6, $\widehat{\psi} = \overline{\psi}$ on $\overline{S}$.

Now define $V^* : \overline{S} \to \mathcal{R}_+$ by $V^*(\overline{\overline{y}}) = \frac{u(\overline{\overline{y}})}{1-\beta}$, and for $s \in [0, \overline{\overline{y}})$

$$\begin{aligned} V^*(s) &= \max_{a \in [0, R_\psi(s)]} \{u(R_\psi(s) - a) + \beta \int V(s')dq(s'|a)\} \\ &= u(R_\psi(s) - \widehat{\psi}(s)) + \beta \int V(s')dq(s'|\widehat{\psi}(s)). \end{aligned} \tag{4.9}$$

We claim that $V^* = V$. To see this note that (by Lemma 2.1 and Theorem 2.1 in [15]) V^* is usc on $\overline{S}$. It is trivial to see that V^* is non-decreasing for a $\in [0, R_\psi(s_1)]$ implies $a \in [0, R_\psi(s_2)]$ whenever $s_1 < s_2$. So clearly $V^* \in \Omega$. As above it suffices to show that $V^* = V$ on D. So let $s \in D$. For each n,

$$V_n(s) = u(R_{\psi_n}(s) - \widehat{\psi}_n(s)) + \beta \int V_n(s')dq(s'|\widehat{\psi}_n(s)) \tag{4.10}$$

and taking limits as $n \to \infty$ yields (since $s \in D$)

$$V(s) = u(R_\psi(s) - \widehat{\psi}(s)) + \beta \int V(s')dq(s'|\widehat{\psi}(s)). \tag{4.11}$$

From (4.9) and (4.11), $V = V^*$ on D, so $V = V^*$ on $\overline{S}$. Thus, we have shown that for $s \in [0, \overline{\overline{y}})$

$$V(s) = \max_{a \in [0, R_\psi(s)]} \{u(R_\psi(s) - a) + \beta \int V(s')dq(s'|a)\} \tag{4.12}$$
$$= u(R_\psi(s) - \widehat{\psi}(s)) + \beta \int V(s')dq(s'|\widehat{\psi}(s)).$$

To complete the proof, it is shown by using similar arguments in [10] and [23] that V is indeed the expected payoff (on $[0, \overline{\overline{y}}]$) from employing the stationary strategy $\widehat{\gamma}(\psi)$ defined by $\widehat{\gamma}(\psi)(y) = y - \gamma(\psi)(y) - \psi(y) = \frac{1}{2}(y + \psi(y)) - \widehat{\psi}(y)$.

Since V satisfies the Bellman Optimality Equation (4.12), and $\widehat{\psi}$ yields a total expected payoff of V, it is indeed the case that $B(\psi) = \widehat{\psi}$. ◇

Combining Lemmas 4.1 and 4.7, we see the existence of a $\psi^* \in \Psi$ such that $B(\psi^*) = \psi^*$. Therefore, there is a function $\widehat{\gamma}^* = \widehat{\gamma}(\psi^*)$, such that $\widehat{\gamma}^*$ is a GBR to itself on $[0, \overline{\overline{y}})$ for problem (P). Denote the restrictions of $\widehat{\gamma}^*$ to S by γ^*.

Lemma 4.8: γ^* is a GBR to itself on S.

Proof: By our assumptions on q, if the game starts with the state in S, the state stays in S forever. If $\widehat{\gamma}^*$ is a GBR to itself on S, then γ^* must be a GBR to itself on S for what happens in $(\overline{y}, \overline{\overline{y}})$ is now irrelevant. ◇

The next two results (finally!) establish the existence of a non-trivial Nash equilibrium in non-randomized stationary strategies to the stochastic game of Section 2.

Lemma 4.9: $2\gamma^*(s) < s$ for all $0 < s \leq \overline{y}$.

Proof: As with Lemma 4.10 below, this result is a simple implication of assumption (U2) that $lim_{x \to 0} u'(x) = \infty$. See [11] for details. ◇

Lemma 4.10: γ^* is a BR to itself on S for the stochastic game of Section 2.

The existence of a non-trivial equilibrium is thus established by Lemmas 4.9 and 4.10. To see that it satisfies the other properties outlined in section 1, note that the ψ^* that, as a fixed-point of B, generated the γ^* is non-decreasing, so for $s_1 \neq s_2 \in S$,

$$\frac{\gamma^*(s_1) - \gamma^*(s_2)}{s_1 - s_2} = \frac{1}{2}\left[\frac{s_1 - s_2 - \psi^*(s_1) + \psi^*(s_2)}{s_1 - s_2}\right]$$
$$= \frac{1}{2}\left[1 - \frac{\psi^*(s_1) - \psi^*(s_2)}{s_1 - s_2}\right]$$
$$\leq \frac{1}{2},$$

and finally since ψ^* is usc on S and γ^* is defined by $\gamma^*(s) = \frac{1}{2}(s - \psi^*(s))$, γ^* is lsc on S. ◇

References

[1] *Bernheim, B.D., and D. Ray* (1987). Economic Growth with Intergenerational Altruism, *Review of Economic Studies* LIV **2**, 227-241.

[2] *Billingsley, P.* (1968). *Convergence of Probability Measures* , Wiley, New York.

[3] *Blackwell, D.* (1965). Discounted Dynamic Programming, *Annals of Mathematical Statistics*, **36**, 226-235.

[4] *Clemhout, S., and H.Y. Wan, Jr.* (1985). Cartelization Conserves Endangered Species? An Application of Phase Diagram to Differential Games, in *Economic Applications of Control Theory II* (G. Feshtinger, ed.). North Holland.

[5] *Debreu, G.* (1952). A Social Equilibrium Existence Theorem, *Proceedings of the National Academy of Sciences, U.S.A.*, **38**, 886-893.

[6] *Debreu, G.* (1959). *Theory of Value*, Yale University Press, New Haven.

[7] *Hildenbrand, W.* (1974). *Core and Equilibria of a Large Economy*, Princeton University Press, Princeton, New Jersey.

[8] *Himmelberg, C., T. Parthasarathy, T.E.S. Raghavan, and F. Van Vleck* (1976). Existence of p-Equilibrium and Optimal Stationary Strategies in Stochastic Games, *Proceedings of the American Mathematical Society*, **60**, 245-261.

[9] *Levhari, D., and L. Mirman* (1980). The Great Fish War: An Example Using a Dynamic Cournot-Nash Solution, *Bell Journal of Economics*, **11**(1), 322-334.

[10] *Maitra, A.* (1968). Discounted Dynamic Programming on Compact Metric Spaces, *Sankhya*, Series A, **30**, 211-221.

[11] *Majumdar, M. and R. Sundaram* (1988). Symmetric Stochastic Games of Resource Extraction: The Existence of non-randomized Stationary Equilibrium, Center for Analytic Economics Working Paper, Department of Economics, Cornell University.

[12] *Mertens, J.-F., and T. Parthasarathy* (1987). Nonzero-Sum Discounted Stochastic Games, private communication from the second author.

[13] *Nowak, A.S.* (1985). Existence of Equilibrium Stationary Strategies on Discounted Non-Cooperative Stochastic Games with Uncountable State Space, *Journal of Optimization Theory and Applications*, **45**, 591-602.

[14] *Parthasarathy, K.R.* (1967). *Probability Measures on Metric Spaces*, Academic Press, New York.

[15] *Parthasarathy, T.* (1973). Discounted, Positive, and Non-Cooperative Stochastic Games, *International Journal of Game Theory*, **2**, 25-37.

[16] *Parthasarathy, T.* (1982). Existence of Equilibrium Stationary Strategies in Discounted Stochastic Games, *Sankhya*, Series A, **44**, 114-127.

[17] *Parthasarathy, T., and S. Sinha* (1987). Existence of Stationary Equilibrium Strategies in Nonzero-Sum Discounted Stochastic Games with Uncountable State Space and State-independent Transition, Technical Report No. 8618, Indian Statistical Institute, New Delhi.

[18] *Ray, D.* (1987). Nonpaternalistic Intergenerational Altruism, *Journal of Economic Theory*, **41** (1). 112-132.

[19] *Rogers, P.D.* (1969). Nonzero-Sum Stochastic Games, Ph.D. Thesis, submitted to the University of California, Berkeley.

[20] *Shapley, L.* (1953). Stochastic Games, *Proceedings of the National Academy of Sciences, U.S.A.*, **39**, 1095-1100.

[21] *Smart, D.* (1974). *Fixed-point Theorems*, Cambridge University Press.

[22] *Sobel, M.J.* (1969). Non-cooperative Stochastic Games, Report No. 21, Department of Administrative Sciences, Yale University.

[23] *Strauch, R.E.* (1966). Negative Dynamic Programming, *Annals of Mathematical Statistics*, **37**, 871-890.

[24] *Sundaram, R.* (1987). Nash Equilibrium in a Class of Symmetric Dynamic Games: An Existence Theorem, Working Paper 87-11, Center for Analytic Economics, Cornell University, Ithaca, New York; revised version forthcoming in *Journal of Economic Theory*.

Mukul Majumdar
Department of Economics
Cornell University
Ithaca, NY 14853
U.S.A.

Rangarajan Sundaram
Department of Economics
University of Rochester
Rochester, NY 14627
U.S.A.

Robustness of Uncertain Dynamical Systems with Delay in the Absence of Matching Assumptions*

Xiong Zhong Kai[1], George Leitmann and Franco Garofalo[1]

Abstract: The theory of uncertain dynamical systems [4,5] provides a useful tool for the design of controllers for systems subject to parameter and disturbance uncertainties when only a range of variation of these uncertainties is known.

The performance guaranteed by these controllers is the ultimate boundedness of the controlled variables. Moreover, increasing the loop gain, the set of ultimate boundedness can be made arbitrarily small. For the design of these controllers, beside the just mentioned information about the possible size of the uncertainties affecting the system, some structural conditions regarding the way in which the uncertainties enter into the system are required. They are the so-called "matching conditions".

Matching of the uncertainties was assumed in an earlier design of the controller of a plant containing delay in the state variables [6]. The aim of this paper is to show that the controller designed in [6] is robust with respect to unmatched uncertainties, provided that they do not exceed an *a priori* determined threshold value. The technique for doing so is similar to the one developed in [7] for the case of plants without time delays.

Introduction: Analysis and control of linear or non-linear dynamical systems with uncertainty provide theoretical as well as practically significant problems. In order to control the behavior of a system in the real world, be it physical, biological, or socioeconomic, the system analyst seeks to capture the system's salient features in a mathematical model. This abstraction of the real system always contains uncertain elements; these may be parameters, constant or varying, which are unknown or imperfectly known, or they may be unknown or imperfectly known inputs into the

*Based, in part on research supported by the NSF.

[1]Visiting scholar, University of California.

Key Words: Uncertain systems, delay systems, mismatched systems.

system, as well as state measurement error. Despite such imperfect knowledge about the chosen mathematical model, one often seeks to devise controllers which will steer the system in some desired fashion, for example so that the system response will approach or track a desired reference response; by suitable definition of system (state) variables, such problems can always be cast into the form of stability problems.

Two main avenues are open to the analyst seeking to control an uncertain dynamical system. He may choose a stochastic approach in which information about the uncertain elements as well as about the system response is statistical in nature; for example, see Refs. [1] and [2]. Loosely speaking, when modeling via random variables, one is content with desirable behavior on the average. The other approach to the control of uncertain systems, and the one for which we shall opt in the present discussion, is deterministic. Available, or assumed, information about uncertain elements is deterministic in nature. Here, one seeks controllers which assure the desired response of the dynamical system.

In this paper the main goal is to study the robustness of a controlled system with delay when the uncertainties do not satisfy the so-called matching conditions. Since no unique definition of robustness of systems currently exists, it is used here to imply satisfactory performance in the presence of unstructured uncertainty. We shall decompose the uncertainty into two parts -- the matched part and the mismatched part. With the previously obtained knowledge that arbitrarily large matched uncertainty can be coped with by certain classes of controllers, the bound of mismatched uncertainty and its maximum allowable size will be studied such that satisfactory system behavior can still be guaranteed. Our principal result is to give a general method for the estimation of this bound and its upper limit. We emphasize here that, even though this bound and its upper limit are dependent on the specific controller employed, this general result is applicable to any admissible control.

2. System Description: Let $C = C([-r,0], R^n)$ be the space of all continuous functions which map the real interval $[-r,0]$ into R^n. The function segment $x_t \triangleq \{x(t+\theta), \theta \in [-r,0]\}$ can be chosen as the state of dynamic systems governed by functional differential equations with delays in the argument of the dependent variable bounded by $r > 0$; C is then a state space for the system. Let $\|\cdot\|$ denote the Euclidean norm and $|\cdot|$ the sup norm on C given by $|x_t| \triangleq sup_{\theta\in[-r,0]} \|x(t+\theta)\|$. The solution at time t of the system equation with initial condition $x_{t_o} = \phi \in C$ is symbolized by $x_t(t_o, \phi)$ in C and by $x(t_o,\phi)(t)$ in R^{n2} With this notation $x_t(t_o,\phi)(\theta) = x(t_o,\phi)(t+\theta)$ for all $\theta \in [-r,0]$, which can be expressed more simply as $x_t(\theta) = x(t+\theta)$ when the initial conditions are clearly understood from the context. Finally, let P and Q be compact subset of R^a and R^b, respectively,

[2]When no confusion is likely to arise we will shorten $x(t_o,\phi)(t)$ to $x(t)$.

and let $R^+ = \{s \in R \mid s \geq 0\}$. The non-linear, time-varying delay system being studied is governed by

$$\dot{x}(t) = f(t, x_t) + \Delta f_1(t, x_t, p(t, x_t)) + [B(t, x_t) + \Delta B_1(t, x_t, q(t, x_t))]u(t) \quad (1)$$
$$+ C_1(t, x_t)v(t)$$

where $x_t \in C$ is the system state and $u(t) \in R^m$ is the control input. The n-dimensional vector $\Delta f_1(t, x_t, p)$ models the uncertainty in the homogeneous part of the system which depends on the uncertainty vector $p \in P$; the $n \times n$ matrix $\Delta B_1(t, x_t, p = q)$ models the uncertainty in the system's control input matrix which depends on uncertainty vector $q \in Q$, and $C_1(t, x_t)v(t)$ models the effect of the input disturbance $v(t) \in R^c$ on $\dot{x}(t)$.

We assume the following decomposition of Δf_1, ΔB_1 and C_1:

$$\begin{aligned} \Delta f_1(t, x_t, p(t, x_t)) &= B(t, x_t)\Delta f_o(t, x_t, p(t, x_t)) + \Delta \tilde{f}_o(t, x_t, p(t, x_t)) \\ \Delta B_1(t, x_t, q(t, x_t)) &= B(t, x_t)\Delta B_o(t, x_t, q(t, x_t)) + \Delta \tilde{B}_o(t, x_t, q(t, x_t)) \quad (2) \\ C_1(t, x_t) &= B(t, x_t)C(t, x_t) + \tilde{C}(t, x_t). \end{aligned}$$

In the general case, the components of the parameter vectors $p(t, x_t)$ and $q(t, x_t)$ are allowed to be continuously dependent on both x_t and t. Then define

$$\begin{aligned} \Delta f(t, x_t) &\triangleq \Delta f_o(t, x_t, p(t, x_t)) \\ \Delta B(t, x_t) &\triangleq \Delta B_o(t, x_t, q(t, x_t)) \quad (3) \\ \Delta \tilde{f}(t, x_t) &\triangleq \Delta \tilde{f}_o(t, x_t, p(t, x_t)) \\ \Delta \tilde{B}(t, x_t) &\triangleq \Delta \tilde{B}_o(t, x_t, q(t, x_t)). \end{aligned}$$

Substitution from (2) and (3) into (1) yields

$$\dot{x}(t) = f(t, x_t) + B(t, x_t)u(t) + B(t, x_t)e(t, x_t) + e(t, x_t) \quad (4)$$

where

$$\begin{aligned} e(t, x_t) &\triangleq \Delta f(t, x_t) + C(t, x_t)v(t) + \Delta B(t, x_t)u(t) \quad (5) \\ e(t, x_t) &\triangleq \Delta f(t, x_t) + \tilde{C}(t, x_t)v(t) + \Delta B(t, x_t)u(t) \end{aligned}$$

The related system

$$\dot{x}(t) = f(t, x_t) + B(t, x_t)u(t) \quad (6)$$

obtained from (4) when $e(t, x_t) \equiv 0$ and $e(t, x_t) \equiv 0$ is called the nominal system.

The related system

$$\dot{x}(t) = f(t, x_t) + B(t, x_t)u(t) + B(t, x_t)e(t, x_t) \quad (6^*)$$

obtained from (4) when $e(t, x_t) \equiv 0$ will be referred to as the system with matched uncertainties.

We now introduce the following definition.

Definition 1: A feed back control $h(\cdot)$: $R \times C \to R^m$ renders the uncertain dynamical system (1) practically stable (p.s.) iff there exist $\underline{d} \in [0, \infty)$ and $r_o > 0$ such that the following properties hold:

(i) Existence of solutions: Given $(t_o, \phi) \in R \times C$, the closed-loop system

$$\dot{x}(t) = f(t, x_t) + B(t, x_t)h(t, x_t) + B(t, x_t)e(t, x_t) + e(t, x_t) \tag{7}$$

possesses a solution $x(\cdot)$: $[t_o, t_1) \to R^n$, $x_{t_o} = \phi$, $t_1 > t_0$.

(ii) Uniform boundedness: Given $\delta \in [0, r_o)$, there exists a positive $d(\delta) < \infty$ such that for all solutions $x(\cdot)$; $[t_o, t_1) \to R^n$, $x_{t_o} = \phi$, of (7), $|\phi| \leq \delta \Rightarrow \|x(t)\| \leq d(\delta) \quad \forall t \in [t_o, t_1)$.

(iii) Extension of solutions: Every solution $x(\cdot)$: $[t_o, t_1) \to R^n$, $x_{t_o} = \phi$, of (7) with $|\phi| \leq \delta$, can be extended over $[t_o, \infty)$.

(iv) Uniform ultimate boundedness: Given any $\epsilon > \underline{d}$ and any $\delta \in [0, r_o)$, there is a $T(\epsilon, \delta) \in [0, \infty)$ such that for every solution $x(\cdot)$: $[t_o, \infty) \to R^n$, $x_{t_o} = \phi$, of (7), $|\phi| < \delta \Rightarrow \|x(t)\| \leq \epsilon \quad \forall t \geq t_o + T(\epsilon, \delta)$.

(v) Uniform stability: Given any $\epsilon > \underline{d}$, there is a positive $\delta(\epsilon) < \infty$ such that for every solution $x(\cdot)$: $[t_o, \infty) \to R^n$, $x_{t_o} = \phi$, of (7),

$$|\phi| \leq \delta(\epsilon) \Rightarrow \|x(t)\| \leq \epsilon \quad \forall t \geq t_0.$$

If $r_o = \infty$ then $h(\cdot)$ renders system (1) globally practically stable (g.p.s).

Definition 2: A functional f: $[t_1, t_2) \times C \to R^n$ is said to be quasi-bounded iff f is bounded on every set $[t_3, t_4] \times C_B$, where $t_1 \leq t_3 < t_4 < t_2$, and $C_b \triangleq C([-r, 0]; B)$, when B is a compact subset of R^n.

The notation of quasi-boundedness is necessary since continuous functionals are not necessarily bounded on closed bounded subsets of $R \times C$.

Definition 3: A functional f: $[t_1, t_2) \times C \to R^n$ is said to be locally Lipschitzian iff, for each $(t, \psi) \in [t_1, t_2) \times C$, there exist positive numbers α and β such that on the set $([t - \alpha, t + \alpha] \cap [t_1, t_2)) \times \{\phi \in C \mid |\phi - \psi| \leq \beta\}$ the functional satisfies $\|f(t, \phi) - f(t, \psi)\| \leq k|\phi - \psi|$ for some $k > 0$.

Concerning the system (1) we introduce the following assumptions.

Assumption 1: The functionals

$$f : R \times C \to R^n, \quad e : R \times C \to R^m,$$
$$B : R \times C \to R^{n \times m}, e : R \times C \to R^n$$

are continuous, quasi-bounded and locally Lipschitzian on $R \times C$.

Assumption 2: There exist a feedback control $h(t, x_t)$, which is continuous, quasi-bounded and locally Lipschitzian on $R \times C$, a continuously differentiable function $V : R \times R^n \to R^+$, continuous, monotone increasing functions $\gamma_1, \gamma_2, \gamma_3, p : R^+ \to R^+$, and a continuous function $\gamma_4 : R^+ \to R$ with the properties $\gamma_1(0) = \gamma_2(0) = \gamma_3(0) = 0$, $lim_{s\to\infty} \gamma_1(s) = \infty$, $lim_{s\to\infty} \gamma_2(s) = \infty$ and $p(s) > s$ for $s > 0$, such that

$$\gamma_1(\|x\|) \leq V(t,x) \leq \gamma_2(\|x\|) \quad \forall t \in R, x \in R^n,$$

and along the solutions of (7)

$$\begin{aligned}\dot{V}(t,x(t)) &\triangleq L_m(t,x(t)) + \nabla_x^T V(t,x(t))e(t,x_t) \\ &\leq -\gamma_3(\|x(t)\|) + \gamma_4(\|x(t)\|) \\ L_m(t,x(t)) &\leq -\gamma_3(\|x\|) \text{ if } V(t+\theta, x(t+\theta)) \\ &\leq p(V(t,x(t)) \quad \forall \theta \in [-r,0]\end{aligned}$$

where

$$L_m(t,x(t)) \triangleq \frac{\partial V(t,x(t))}{\partial t} + \nabla_x^T V(t,x(t))[f(t,x_t) + B(t,x_t)h(t,x_t) + B(t,x_t)e(t,x_t)].$$

Assumption 3: There exist $k_o \in [0,\infty)$ and $M > 0$ (possibly infinite) such that $\gamma_1(M) > \gamma_2(k_o)$ and $\gamma_5(\|x\|) \triangleq \gamma_3(\|x\|) - \gamma_4(\|x\|)$ is non-negative and strictly increasing for $\|x\| \in [k_o, M)$.

3. Main Result: Now we are ready to state the following theorem.

Theorem: Consider the dynamical system (4) subject to feedback control $h(t, x_t)$ and satisfying Assumptions 1-3. Then $h(t, x_t)$ ensures practical stability of the mismatched delay system with $r_o \cdot \underline{d}$, $d(\delta)$ and $T(\epsilon, \delta)$ given by $r_o = (\gamma_2^{-1} \circ \gamma_1)(M)$, $d(\delta) = (\gamma_1^{-1} \circ \gamma_2)(H)$ where $H \triangleq max\{k_o, \delta\}$, $\underline{d} = (\gamma_1^{-1} \circ \gamma_2)(k_o)$, $T(\epsilon, \delta) = N(\frac{\gamma_2(M)}{\gamma}) + (N-1)\gamma$ where $\gamma \triangleq (\gamma_5 \circ \gamma_2^{-1} \circ \gamma_1)(\epsilon)$, and N is the smallest positive integer such that $\gamma_1(\epsilon) + Na \geq \gamma_2(M)$ where $a \triangleq min_s[p(s) - s]$ for $s \in [\gamma_1(\epsilon), \gamma_2(M)]$.

Proof:

(i) (Existence of solutions)

Given any $t_o \in R$ and $\phi \in C$, there exists ([3], p. 301) a unique solution $x(t_o, \phi)(t)$ of the system

$$\dot{x}(t) = f(t, x_t) + B(t, x_t)[h(t, x_t) + e(t, x_t)] + e(t, x_t)$$
$$x_{t_o} = \phi$$

on an interval $[t_o, t_0 + \Delta$, where $\Delta > 0$. We will show next that the existence and uniqueness of every solution $x(t_o, \phi)$ can be extended to the interval $[t_o, \infty)$. For this purpose, it is sufficient to prove that the solutions are uniformly bounded.

(ii) (Uniform boundedness)

Now let the initial function $\phi \in C$ at time t_o satisfy $|\phi| < \delta$ and define $H \triangleq max\{\delta, k_o\}$. Since $lim_{s\to\infty} \gamma_1(s) = \infty$, we can choose $M > H$ such that $\gamma_1(M) > \gamma_2(H)$. We shall show that $\|x(t)\|$ is uniformly bounded by M on $[t_o, t_1]$ where $t_1 > t_o$. Suppose this were not true; then $\|x(t)\| > M$ for some $t \in [t_o, t_1]$. We define $t_2 \triangleq inf\{t \in [t_o, t_1] \mid \|x(t)\| = M\}$ and $t_3 \triangleq sup\{t \in [t_o, t_2] \mid \|x(t)\| = H\}$. By continuity of $x(\cdot)$ and $\|x(t_o)\| \leq H$, t_2 and t_3 are well-defined and, by definition of t_2, $x(t_2) \geq x(t_2 + \theta)$, $\theta \in [-r, 0]$. Hence, if $V(t + \theta, x(t + \theta)) \leq p(V(t, x(t))$,

$$\gamma_1(\|x(t_2)\|) \leq V(t_2, x(t_2)) = V(t_3, x(t_3)) + \int_{t_3}^{t_2} \dot{V}(t, x(t))dt \leq V(t_3, x(t_3))$$
$$+ \int_{t_3}^{t_2} [-\gamma_3(H) + \gamma_4(H)]dt \leq V(t_3, x(t_3)) \leq \gamma_2(H) < \gamma_1(M)$$

from which we obtain the contradiction $\|x(t_2)\| < M$. Consequently $\|x(t)\| \leq M$ on any interval $[t_o, t_1]$, and

(iii) by the extended existence theorem in ([3], p. 306) the unique solution can be continued on all of $[t_o, \infty)$.

(iv) (Uniform ultimate boundedness)

Let $x(\cdot) : [t_0, \infty) \to R^n$, $x_{t_o} = \phi$, be a solution with given $|\phi| < \delta$. Then as shown above, $\|x(t)\| \leq M$ and $V(t, x(t)) \leq \gamma_2(M)$ for $t \geq t_o - r$. Choose ϵ such that $0 \leq (\gamma_1^{-1} \circ \gamma_2)(k_o) < \epsilon$. We want to show that $\|x(t)\| \leq \epsilon$ for all $t \geq t_o + T$, where T is said to be determined (T is independent of t_o, but may depend on ϵ and δ). If $\epsilon \geq M$, $\|x(t)\| \leq \epsilon$ for $t \geq t_o - r$ follows directly from the uniform boundedness result (ii). So we consider $\epsilon < M$. Define $a \triangleq min_s[p(s) - s]$, $\gamma_1(\epsilon) \leq s \leq \gamma_2(M)$. Let N be the smallest positive integer such that $\gamma_1(\epsilon) + Na \geq \gamma_2(M)$, and let $\gamma \triangleq (\gamma_5 \circ \gamma_2^{-1} \circ \gamma_1)(\epsilon) > 0$. We want to show $V(t, x(t)) \leq \gamma_1(\epsilon) + (N-1)a$ for $[t_o, t_o + T_1]$, $T_1 = \frac{\gamma_2(M)}{\gamma}$. Suppose, on the contrary, that $\gamma_1(\epsilon) + (N-1)a < V(t, x(t))$ for any $t > t_o$; then

$$p(V(t, x(t)) \geq V(t, x(t)+a > \gamma_1(\epsilon)+Na \geq \gamma_2(M) \geq V(t+\theta, x(t+\theta)), \quad \theta \in [-r, 0]$$

and hence $\dot{V}(t, x(t)) \leq -\gamma_5(\|x(t)\|) \leq -(\gamma_5 \circ \gamma_2^{-1} \circ \gamma_1)(\epsilon))) \leq -\gamma$ at any such $t > t_o$. Note that $\|x(t)\| < (\gamma_2^{-1} \circ \gamma_1)(\epsilon)$ leads to the contradiction $\gamma_2(\|x(t)\|) < \gamma_1(\epsilon) < V(t, x(t)) \leq \gamma_2(\|x(t)\|)$.

Since $V(t, x(t)) \leq V(t_o(x(t_o)) - \gamma(t-t_o) \leq \gamma_2(M) - \gamma(t-t_o)$ and $V(t, x(t)) \geq 0$, we must have $V(t, x(t)) \leq \gamma_1(\epsilon) + (N-1)a$ for some $[t_o, t_o+T_1]$, where $T_1 \triangleq \frac{\gamma_2(M)}{\gamma}$. However, if $V(t, x(t)) = \gamma_1(\epsilon) + (N-1)a$ then $p(V(t, x(t))) \geq V(t, x(t)) + a \geq \gamma_1(\epsilon) + Na \geq \gamma_2(M) \geq V(t+\theta, x(t+\theta))$, $\theta \in [-r, 0]$; so $\dot{V} \leq -\gamma < 0$. Hence, $V(t, x(t)) \leq \gamma_1(\epsilon) + (N-1)a$ for all $t \geq t_0 + T_1$. If $N = 1$, the desired result $\|x(t)\| \leq \epsilon$ for all $t \geq t_o + T$ (with $T \triangleq T_1$) then follow directly from $\gamma_1(\|x(t)\|) \leq V(t, x(t)) \leq \gamma_1(\epsilon) + (N-1)a$. For the case $N > 1$, we proceed by mathematical induction. For $k = 1, 2, \ldots, N$, define $T_k = k(\frac{\gamma_2(M)}{\gamma})$, and let $V(t, x(t)) \leq \gamma_1(\epsilon) + (N-k)a$ for all $t \geq t_o + T_k + (k-1)\gamma$. If $\gamma_1(\epsilon) + [N - (k+1)]a < V(t, x(t))$ for $t \geq t_o + T_k + k_\gamma$, then $p(V(t, x(t)) \geq V(t, x(t)) + a \geq \gamma_1(\epsilon) + (N-k)a \geq V(t+\theta, x(t+\theta))$, $\theta \in [-r, 0]$, so $\dot{V} \leq -\gamma$. Arguments similar to those employed in the interval $[t_o, t_o + T_1]$ above then prove that $V(t, x(t)) \leq \gamma_1(\epsilon) + (N - (k+1))a$ for all $t \geq t_o + T_{k+1} + k\gamma$. Hence,

$\gamma_1(\|x(t)\|) \leq V(t, x(t)) \leq \gamma_1(\epsilon)$ for all $t \geq t_o + T$, where $T \triangleq T_N + (N-1)\gamma$.

(v) (Uniform stability)

Let $\epsilon > \underline{d} \triangleq (\gamma_1^{-1} \circ \gamma_2)(k_o)$ be specified. We take $\delta(\epsilon) = (\gamma_2^{-1} \circ \gamma_1)(\epsilon)$ for $\epsilon \in (\underline{d}, M)$ and $\delta(\epsilon) \leq (\gamma_2^{-1} \circ \gamma_1)(M)$ for $\epsilon \geq M$. Then, given any solution $x(t) : [t_o, \infty) \to R^n$ with $\|\phi\| \leq \delta(\epsilon)$, it follows from (ii) and (iv) that $\|x(t)\| \leq \epsilon$ for all $t \geq t_o$. Figure 1 illustrates several related quantities of the Theorem.

Example: To illustrate the preceding result, consider the mismatched delay system

$$\dot{x}(t) = a(t)x(t) + b(t)x(t-\tau) + c(t)u(t) + \Delta c(t) + d(t)$$

with

$$a(t) = 12\,cos\,t - 0.5, b(t) = (1 + 0.5\,sin\,t), c(t) = 0.5 + cos\,t,$$

$$\Delta c(t) = (0.5 + cos\,t)x(t)\Delta c(t); 0 < \tau \leq r;$$

$$\|d(t)\| \leq 1; (0.5 + cos\,t)\Delta c(t) \leq 0.5 + cos\,t.$$

If we use the feedback law $u(t) = -10x(t)$, we can asymptotically stabilize the matched system

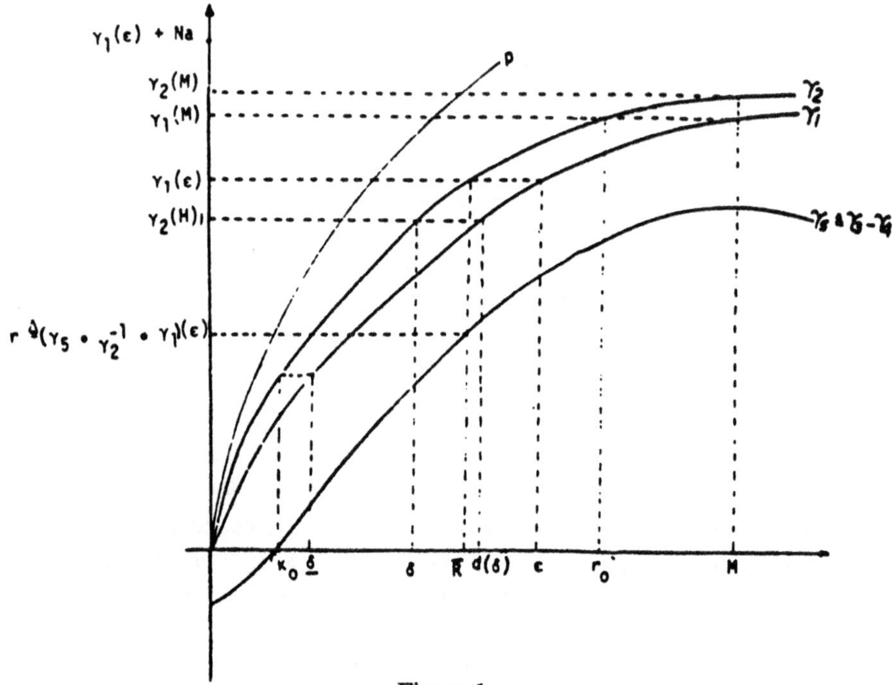

Figure 1
Several related quantities for Theorem

$$\dot{x}(t) = (12\,cos\,t - 0.5)x(t) + (1 + 0.5\,sin\,t)x(t-r) + (0.5 + cos\,t)(-10x(t)) + (0.5 + cos\,t)x(t)\Delta c(t).$$

Indeed, with $V(x) = \frac{1}{2}x^2$ and $p(s) = 1.44s$,

$$\begin{aligned} L_m(t, x_t) &= x(t) \cdot \dot{x}(t) = (12\,cos\,t - 0.5)x^2(t) + (1 + 0.5\,sin\,t)x(t)(t-r) \\ &\quad + [-5x^2(t) - 10x^2(t)\,cos\,t] + (0.5 + cos\,t)x^2(t)\Delta c(t) \\ &\leq (12\,cos\,t - 0.5)x^2(t) + (1 + 0.5\,sin\,t)x(t)x(t-r) \\ &\quad + [-5x^2(t) - 10x^2(t)\,cos\,t] + (0.5 + cos\,t)x^2(t) \\ &= -(5 - 3\,cos\,t)x^2(t) + (1 + 0.5\,sin\,t)x(t)x(t-r) \\ &\leq -0.2x^2(t) \triangleq -\gamma_3(\|x(t)\|) \end{aligned}$$

whenever $p(V(x(t))) \geq V(x(t+\theta))$, $\theta \in [-r, 0]$. Assumption 3 is then satisfied with $\gamma_3(s) = 0.2s^2$ and $\gamma_1(s) = \gamma_2(s) = \frac{1}{2}s^2$. Furthermore,

$$\nabla^T V e = x \cdot d(t) \leq \|x\| \stackrel{\Delta}{=} \gamma_4(\|x\|)$$

and

$$\gamma_5(\|x\|) = \gamma(\|x\|) - \gamma_4(\|x\|) = 0.2x^2 - \|x\|,$$
$$\dot{V}(t, x(t)) = L_m(t, x_t) + \nabla^T V(t, x_t) e(t, x_t) \leq -\gamma_5(\|x(t)\|) = -0.2x^2(t) + \|x(t)\|$$

whenever $p(V(x(t))) \geq V(x(t+\theta))$, $\theta \in [-r, 0]$. Hence, $M = \infty$, $k_o = 5$, $\gamma_5(\|x\|) \geq 0 \quad \forall \|x\| \in [5, \infty)$. If $\delta = 0,5$, $H = max\{k_o, \delta\} = 5$, $M > \epsilon = 8 > (\gamma_1^{-1} \circ \gamma_2)(k_o) = k_o = 5$, $a = min_{s \geq \gamma_1(\epsilon)}[p(s) - s] = min_{s \geq 32}[1.44s - s] = 14.08$, $\gamma = (\gamma_5 \circ \gamma_2^{1-} \circ \gamma_1)(8) = \gamma_5(8) = 4.8$. If we let $M = 10$, then $N = 2$,

$$T(\epsilon, \delta) = N\left(\frac{\gamma_2(M)}{\gamma}\right) + (N-1)\gamma \geq 26, d(\delta) = (\gamma_1^{-1} \circ \gamma_2)(H) = 5,$$
$$r_o = (\gamma_2^{-1} \circ \gamma_1)(M) = 10, \underline{d} = (\gamma_1^{-1} \circ \gamma_2)(k_o) = 5.$$

5. Conclusions: It has been shown in this paper that practical stability of mismatched non-linear delay systems can be obtained under certain specific conditions.

References

[1] *Kushner, J.J.* (1966). On the Status of Optimal Control and Stability for Stochastic Systems, *IEEE International Convention Rec.*, **14**, 143-151.

[2] *Aström, K.J.* (1970). Introduction to Stochastic Control Theory, Academic Press, New York.

[3] *Driver, R.D.* (1977). Ordinary and Delay Differential Equations, Springer-Verlag, New York.

[4] *Corless, M.J. and G. Leitmann* (1981). *IEEE Trans. Autom. Control*, **26**, 1139-1144.

[5] *Leitmann, G.* (1978). *IEEE Trans. Autom. Control*, **23**, 1109-1110; (1979). *J. Dynamic System Meas. Control*, **101**, 212-216; (1981). *J. Dynamic System Meas. Control*, **103**, 95-102.

[6] *Thowsen, A.* (1983). *Int. J. Control*, **3**, No. 5, 1135-1143.

[7] *Chen, Y.H. and Leitmann, G.* (1987). *Int. J. Control*, **45**, 1527-1542.

[8] *Chen, Y.H.* (1987). *ASME J. Dynamic Systems Meas. Control*, **109**, 29-35.

Xiong Zhong Kai
Chongqing Communication Institute
Chongqing, Sichuan
People's Republic of China

G. Leitmann
University of California
Berkeley, California
U.S.A.

F. Garofalo
Dipartimento di Informatica e Sistemistica
Universita di Napoli
via Claudio 21
80127 Napoli
Italy

Stochastic Games[1,2]

L.S. Shapley[3]

Introduction. -- In a stochastic game the play proceeds by steps from position to position, according to transition probabilities controlled jointly by the two players. We shall assume a finite number, N, of positions, and finite numbers m_k, n_k of choices at each position; nevertheless, the game may not be bounded in length. If, when at position k, the players choose their ith and jth alternatives, respectively, then with probability $s_{ij}^k > 0$ the game stops, while with probability p_{ij}^{kl} the game moves to position l. Define

$$s = \min_{k,i,j} s_{ij}^k.$$

Since s is positive, the game ends with probability 1 after a finite number of steps, because, for any number t, the probability that it has *not* stopped after t steps is not more than $(1-s)^t$.

Payments accumulate throughout the course of play: the first player takes a_{ij}^k from the second whenever the pair i, j is chosen at position k.
If we define the bound M:

$$M = \max_{k,i,j} |a_{ij}^k|,$$

then we see that the expected total gain or loss is bounded by

$$M + (1-s)M + (1-s)^2 M + \ldots = M/s. \tag{1}$$

The process therefore depends on $N^2 + N$ matrices

$$P^{kl} = (p_{ij}^{kl} | i = 1, 2, \ldots, m_k;\ j = 1, 2, \ldots, n_k)$$

---

[1]The preparation of this paper was sponsored (in part) by the Office of Naval Research.

[2]Communicated by J. von Neumann, July 17, 1953.

[3]Princeton University.

$$A^k = (a_{ij}^k | i = 1, 2, \ldots, m_k; \; j = 1, 2, \ldots, n_k),$$

with $k, l = 1, 2, \ldots, N$, with elements satisfying

$$p_{ij}^{kl} \geq 0, \; |a_{ij}^k| \leq M, \; \sum_{l=1}^{N} p_{ij}^{kl} = 1 - s_{ij}^k \leq 1 - s < 1.$$

By specifying a starting position we obtain a particular game Γ^k. The term "stochastic game" will refer to the collection $\Gamma = \{\Gamma^k | k = 1, 2, \ldots, N\}$.

The full sets of pure and mixed strategies in these games are rather cumbersome, since they take account of much information that turns out to be irrelevant. However, we shall have to introduce a notation only for certain behavior strategies[1] , namely those which prescribe for a player the same probabilities for his choices every time the same position is reached, by whatever route. Such *stationary strategies*, as we shall call them, can be represented by N-tuples of probability distributions, thus:

$$\vec{x} = (x^1, x^2, \ldots, x^N), \quad \text{each } x^k = (x_1^k, x_2^k, \ldots, x_{m_k}^k),$$

for the first player, and similarly for the second player. This notation applies without change in all of the games belonging to Γ.

Note that a stationary strategy is not in general a mixture of *pure* stationary strategies (all x_i^k zero or one), since the probabilities in a behavior strategy must be uncorrelated.

Existence of a solution. -- Given a matrix game B, let val $[B]$ denote its minimax value to the first player, and $X[B]$, $Y[B]$ the sets of optimal mixed strategies for the first and second players, respectively[2] . If B and C are two matrices of the same size, then it is easily shown that

$$|\text{ val } [B] - \text{ val } [C]| \leq \max_{i,j} |b_{ij} - c_{ij}|. \tag{2}$$

Returning to the stochastic game Γ, define $A^k(\vec{\alpha})$ to be the matrix of elements

$$a_{ij}^k + \textstyle\sum_l p_{ij}^{kl} \alpha^l,$$

$i = 1, 2, \ldots, m_k; \; j = 1, 2, \ldots, n_k$, where $\vec{\alpha}$ is any N-vector with numerical components. Pick $\vec{\alpha}_{(0)}$ arbitrarily, and define $\vec{\alpha}_{(t)}$ by the recursion:

$$\alpha_{(t)}^k = \text{ val } [A^k(\vec{\alpha}_{(t-1)})], \quad t = 1, 2, \ldots.$$

(If we had chosen $\alpha_{(0)}^k$ to be the value of A^k, for each k, then $\alpha_{(t)}^k$ would be the value of the truncated game $\Gamma_{(t)}^k$ which starts at position k, and which is cut off after t steps if it lasts that long.) We shall show that the limit of $\vec{\alpha}_{(t)}$ as $t \to \infty$ exists and is independent of $\vec{\alpha}_{(0)}$, and that its components are the values of the infinite games Γ^k.

Consider the transformation T:

$$T\vec{\alpha} = \vec{\beta}, \qquad \text{where } \beta^k = \text{ val } [A^k(\vec{\alpha})].$$

Define the norm of $\vec{\alpha}$ to be

$$||\vec{\alpha}|| = \max_k |\alpha^k|.$$

Then we have

$$\begin{aligned} ||T\vec{\beta} - T\vec{\alpha}|| &= \max_k |\text{ val } [A^k(\vec{\beta})] - \text{ val } [A^k(\vec{\alpha})]|| \\ &\leq \max_{k,i,j} |\textstyle\sum_l p_{ij}^{kl}\beta^l - \sum_l p_{ij}^{kl}\alpha^l \\ &\leq \max_{k,i,j} |\textstyle\sum_l p_{ij}^{kl}| \max_l |\beta^l - \alpha^l| \\ &= (1-s)||\vec{\beta} - \vec{\alpha}||, \end{aligned} \tag{3}$$

using (2). In particular, $||T^2\vec{\alpha} - T\vec{\alpha}|| \leq (1-s)||T\vec{\alpha} - \vec{\alpha}||$. Hence the sequence $\vec{\alpha}_{(0)}, T\vec{\alpha}_{(0)}, T^2\vec{\alpha}_{(0)}, \ldots$ is convergent. The limit vector $\vec{\phi}$ has the property $\vec{\phi} = T\vec{\phi}$. Bur there is only one such vector, for $\vec{\psi} = T\vec{\psi}$ implies

$$||\vec{\psi} - \vec{\phi}|| = ||T\vec{\psi} - T\vec{\phi}|| \leq (1-s)||\vec{\psi} - \vec{\phi}||,$$

by (3), whence $||\vec{\psi} - \vec{\phi}|| = 0$. Hence $\vec{\phi}$ is the unique fixed point of T and is independent of $\vec{\alpha}_{(0)}$.

To show that ϕ^k is the value of the game Γ^k, we observe that by following an optimal strategy of the finite game $\Gamma^k_{(t)}$ for the first t steps and playing arbitrarily thereafter, the first player can assure himself an amount within $\varepsilon_t = (1-s)^t M/s$ of the value of $\Gamma^k_{(t)}$; likewise for the other player. Since $\varepsilon_t \to 0$ and the value of $\Gamma^k_{(t)}$ converges to ϕ^k, we conclude that ϕ^k is indeed the value of Γ^k. Summing up:

THEOREM 1. *The value of the stochastic game Γ is the unique solution $\vec{\phi}$ of the system*

$$\phi^k = \text{ val } [A^k(\vec{\phi})], \qquad k = 1, 2, \ldots, N.$$

Our next objective is to prove the existence of optimal strategies.

THEOREM 2. *The stationary strategies $\vec{x}^*, \vec{y}^*$, where $x^l \in X[A^l(\vec{\phi})]$, $y^l \in Y[A^l(\vec{\phi})]$, $l = 1, 2, \ldots, N$, are optimal for the first and second players respectively in every game Γ^k belonging to Γ.*

Proof: Let a finite version of Γ^k be defined by agreeing that on the tth step the play shall stop, with the first player receiving the amount $a_{ij}^h + \sum_l p_{ij}^{hl}\phi^l$ instead of just a_{ij}^h. Clearly, the stationary strategy $\vec{x}^*$ assures the first player the amount ϕ^k in this finite version. In the original game Γ^k, if the first player uses $\vec{x}^*$, his expected winnings after t steps will be at least

$$\phi^k - (1-s)^{t-1} \max_{h,i,j} \textstyle\sum_l p_{ij}^{hl} \phi^l,$$

and hence at least

$$\phi^k - (1-s)^t \max_l \phi^l.$$

His total expected winnings are therefore at least

$$\phi^k - (1-s)^t \max_l \phi^l - (1-s)^t M/s.$$

Since this is true for arbitrarily large values of t, it follows that $\vec{x}^*$ is optimal in Γ^k for the first player. Similarly, $\vec{y}^*$ is optimal for the second player.

Reduction to a Finite-Dimensional Game. -- The non-linearity of the "val" operator often makes it difficult to obtain exact solutions by means of Theorems 1 and 2. It therefore becomes desirable to express the payoff directly in terms of stationary strategies. Let $\vec{\Gamma} = \{\vec{\Gamma}^k\}$ denote the collection of games whose *pure* strategies are the stationary strategies of Γ. Their payoff functions $\mathcal{D}^k(\vec{x}, \vec{y})$ must satisfy

$$\mathcal{D}^k(\vec{x}, \vec{y}) = x^k A^k y^k + \textstyle\sum_l x^k P^{kl} y^k \mathcal{D}^l(\vec{x}, \vec{y}),$$

for $k = 1, 2, \ldots, N$. This system has a unique solution; indeed, for the linear transformation $T_{\vec{x}\ \vec{y}}$:

$$T_{\vec{x}\ \vec{y}} \vec{\alpha} = \vec{\beta}, \quad \text{where } \beta^k = x^k A^k y^k + \textstyle\sum_l P^{kl} y^k \alpha^l$$

we have at once

$$||T_{\vec{x}\ \vec{y}} \vec{\beta} - T_{\vec{x}\ \vec{y}} \vec{\alpha}|| = \max_k |\textstyle\sum_l x^k P^{kl} y^k (\beta^l - \alpha^l)| \leq (1-s) ||\vec{\beta} - \vec{\alpha}||,$$

corresponding to (3) above. Hence, by Cramer's rule,

$$\mathcal{D}^k(\vec{x}, \vec{y}) = \frac{\begin{vmatrix} x^1 P^{11} y^1 - 1 & x^1 P^{12} y^1 & \ldots & -x^1 A^1 y^1 & \ldots & x^1 P^{1N} y^1 \\ x^2 P^{21} y^2 & x^2 P^{22} y^2 - 1 & & & & \\ \ldots & & & \ldots & & \ldots \\ x^N P^{N1} y^N & \ldots & \ldots & -x^N A^N y^N & \ldots & x^N P^{NN} y^N - 1 \end{vmatrix}}{\begin{vmatrix} x^1 P^{11} y^1 - 1 & x^1 P^{12} y^1 & \ldots & x^1 P^{1k} y^1 & \ldots & x^1 P^{1N} y^1 \\ x^2 P^{21} y^2 & x^2 P^{22} y^2 - 1 & & \ldots & & \\ \ldots & & & x^k P^{kk} y^k - 1 & & \ldots \\ & & & \ldots & & \\ x^N P^{N1} y^N & \ldots & \ldots & x^N P^{Nk} y^N & \ldots & x^N P^{NN} y^N - 1 \end{vmatrix}}$$

THEOREM 3. *The games Γ^k possess saddle points:*

$$\min_{\vec{y}} \max_{\vec{x}} \mathcal{D}^k(\vec{x}, \vec{y}) = \max_{\vec{x}} \min_{\vec{y}} \mathcal{D}^k(\vec{x}, \vec{y}), \tag{4}$$

for $k = 1, 2, \ldots, N$. Any stationary strategy which is optimal for all $\Gamma^k \in \Gamma$ is an optimal pure strategy for all $\vec{\Gamma}^k \in \vec{\Gamma}$, and conversely. The value vectors of Γ and $\vec{\Gamma}$ are the same.

The proof is a simple argument based on Theorem 2. It should be pointed out that a strategy $\vec{x}$ may be optimal for one game Γ^k (or $\vec{\Gamma}^k$) and not optimal for other games belonging to Γ (or $\vec{\Gamma}$). This is due to the possibility that Γ might be " disconnected"; however if none of the p_{ij}^{kl} are zero this possibility does not arise.

It can be shown that the sets of optimal stationary strategies for Γ are closed, convex polyhedra. A stochastic game with rational coefficients does not necessarily have a rational value. Thus, unlike the minimax theorem for bilinear forms, the equation (4) is not valid in an arbitrary ordered field.

Examples and Applications. -- 1. When $N-1$, Γ may be described as a simple matrix game A which is to be replayed according to probabilities that depend on the players' choices. The payoff function of Γ is

$$\mathcal{D}(x, y) = \frac{xAy}{xSy},$$

where S is the matrix of (non-zero) stop probabilities. The minimax theorem (4) for rational forms of this sort was established by von Neumann[3] ; an elementary proof was subsequently given by Loomis[4] .

2. By setting all the stop probabilities s_{ij}^k equal to $s > 0$, we obtain a model of an indefinitely continuing game in which future payments are discounted by a factor $(1-s)^t$. In this interpretation the actual transition probabilities are $q_{ij}^{kl} = p_{ij}^{kl}/(1-s)$. By holding the q_{ij}^{kl} fixed and varying s, we can study the influence of interest rate on the optimal strategies.

3. A stochastic game does not have perfect information, but is rather a "simultaneous game", in the sense of Kuhn and Thompson[1] . However, perfect information can be simulated within our framework by putting either m_k or n_k equal to 1, for all values of k. Such a stochastic game of perfect information will of course have a solution in stationary *pure* strategies.

4. If we set $n_k = 1$ for all k, effectively eliminating the second player, the result is a "dynamic programming" model[5] . Its solution is given by any set of integers $\vec{i} = \{i_1, i_2, \ldots, i_N | 1 \leq i_k \leq m_k\}$ which maximizes the expression

$$\mathcal{A}^k(i) = \frac{\begin{vmatrix} p_{i_1}^{11}-1 & p_{i_1}^{12} & \cdots & -a_{i_1}^1 & \cdots & p_{i_1}^{1N} \\ p_{i_2}^{21} & p_{i_2}^{22}-1 & & & & \\ \cdots & & & \cdots & & \cdots \\ p_{i_N}^{N1} & & \cdots & -a_{i_N}^N & \cdots & p_{i_N}^{NN}-1 \end{vmatrix}}{\begin{vmatrix} p_{i_1}^{11}-1 & p_{i_1}^{12} & \cdots & p_{i_1}^{1k} & \cdots & p_{i_1}^{1N} \\ p_{i_2}^{21} & p_{i_2}^{22}-1 & & \cdots & & \\ \cdots & & & p_{i_k}^{kk}-1 & & \cdots \\ & & & \cdots & & \\ p_{i_N}^{N1} & & \cdots & p_{i_N}^{Nk} & \cdots & p_{i_N}^{NN}-1 \end{vmatrix}}$$

For example (taking $N = 1$), let there be alternative procedures $i = 1, \ldots, m$ costing $c_i = -a_i$ to apply and having probability s_i of success. The above then gives us the rule: adopt that procedure i^* which maximizes the ration a_{i*}/s_{i*}, or equivalently, the ratio s_{i*}/c_{i*}.

5. Generalizations of the foregoing theory to infinite sets of alternatives, or to an infinite number of states, readily suggest themselves (see for example ref. 6). We shall discuss them in another place.

[1]Kuhn, H.W., *Contributions to the Theory of Games II*, Annals of Mathematics Studies No. 28, Princeton, 1953, pp. 209-210.

[2]Von Neumann, J., and Morgenstern, O., *Theory of Games and Economic Behavior*, Princeton, 1944 and 1947, p. 158.

[3]Von Neumann, J., *Ergebnisse eines Math. Kolloquiums*, 8, 73-83 (1937).

[4]Loomis, L.H., these PROCEEDINGS, 32, 213-215 (1946).

[5]Bellman, R., these PROCEEDINGS, 38, 716-719 (1952).

[6]Isbell, J.R., *Bull. A.M.S.*, 59, 234-235 (1953).

On Games of Survival[1]

J. Milnor and L.S. Shapley

Two gamblers, with limited resources of money, agree to play and replay the same zero-sum game until one of them is ruined. The "game of survival" that results is similar in many respects to the classic "gambler's ruin" problem, but there is one important difference: since the transition probabilities are controlled by the participants, and not by chance, there may be a positive probability of infinite repetition, with neither gambler being ruined. Thus, to save oneself and to destroy one's opponent are somewhat different objectives; in fact , the optimal strategies and the corresponding probabilities of ruin or survival wil sometimes be found to depend on the value assigned to the case of double survival.

In this paper we propose to investigate thoroughly those games of survival where the underlying "money" game is given by an arbitrary, finite matrix of real numbers. The existence of solutions, and the extent to which they depend on the double-survival payoff, are the central topics. Our approach combines an analysis of certain game-theoretic functional equations with the theory of semimartingales. A number of examples, and methods of constructing and approximating the solutions, are also discussed.

Previous writings on the subject include those of Bellman and LaSalle [3], Hausner [10], Peisakoff [11], and Bellman [1, 2]; however, only certain special cases have so far been examined rigorously. Usually, it has been assumed that the number of accessible "states" (possible distributions of money during the play) is finite. It

[1]Portions of this paper were presented by the authors at a conference on "Recent Developments in the Theory of Games" in Princeton, January 31 - February 1, 1955, and at a meeting of the Mathematical Association of America in Seattle, August 20-21, 1956. The work was supported by the RAND Corporation and by the Office of Naval Research through contract Nonr-220(16) with the California Institute of Technology.

should be remarked that under such restructions survival games become "recursive games" in the sense of Everett, and the existence of a solution (assuming a constant double- survival payoff) becomes a simple corollary of his result [6]. Mention should also be made of the multi-dimensional survival games treated by Scarf [12], and the somewhat similar multi-component attrition games of Blackwell [4].

§1. GENERAL INTRODUCTION AND EXAMPLES

Let $||a_{ij}||$ denote the matrix of the "money" game, let R be the sum of the resources of the two players, and let r_0 be the first player's initial fortune. Then, if player I chooses i_k on the kth round and Player II chooses j_k, the new level of Player I's fortune is given by:

(1) $$r_k = r_{k-1} + a_{i_k j_k}, \quad k = 1, 2, \ldots ;$$

a formula valid so long as $0 < r_{k-1} < R$. For r_{k-1} outside this interval we define $r_k = r_{k-1}$, serving the formal purpose of associating an infinite sequence $\{r_k\}$ with every play of the game, whether it terminates or not.

If one of the players is eventually ruined, the "utility" payoff to Player I can be defined:

(2) $$P(r) = \begin{cases} 0 & \text{if} \quad r \leq 0 \\ 1 & \text{if} \quad r \geq R \end{cases},$$

where $r = lim\ r_k$. If both players survive indefinitely, the payoff will be a number Q, which may be a function of the course of play. When we are not considering special cases we shall let Q be entirely arbitrary, assuming only $0 \leq Q \leq 1$. The payoff to Player II is taken to be 1 minus the payoff to Player I.

Thus, the survival game is completely specified by the four elements $||a_{ij}||$, r_0, R, and Q; together with the specified payoff function P.

Assume for the moment that Q is a constant, and that the value of the game exists for every initial state r_0. Then it is easily proved that the value is a monotonic increasing function of r_0, and that it satisfies the functional equation:

(3) $$\phi(r) = \text{val}||\phi(r + a_{ij})||, \quad 0 < r < R,$$

with boundary conditions:

(4) $$\phi(r) = P(r) \quad r \leq 0, \quad r \geq R.$$

Here "val" denotes the ordinary minimax value of a matrix game.

Even if Q is not constant, equations (3) and (4) play a very fundamental role in the analysis. As we shall see in Section 2 of this paper, there always exists at least one monotonic solution to (3), (4).

If this solution is unique, then the value of the survival game exists and is independent of Q. If the solution is not unique, then the value may not exist, and it is not independent of Q if it does exist.

[To illustrate: in the first example below, all monotonic functions, and some others, are solutions of (3). In the second example, any linear or near-linear function is a solution. In both cases the dependence of the game on Q is intuitively obvious, since both players have powerful "defensive" strategies that prevent any action from taking place except on favorable ground. In the third

$$\begin{pmatrix} 0 & 1 \\ -1 & 0 \end{pmatrix} \qquad \begin{pmatrix} 0 & 0 & 2 & -1 \\ 0 & 0 & -1 & 2 \\ 1 & -2 & 0 & 0 \\ -2 & 1 & 0 & 0 \end{pmatrix} \qquad \begin{pmatrix} 0 & 0 \\ 0 & 0 \end{pmatrix}$$

Example 1 Example 2 Example 3

example equation (3) becomes trivial and irrelevant. The game turns entirely on the properties of Q. An example of Gale and Stewart [8] shows that the value does not exist for certain payoffs of the form $Q = Q(i_1, j_2, i_3, \ldots)$. Whether such indeterminacy can ever occur when Q has the form $Q(r_1, r_2, \ldots)$ is an open question.]

A mixed strategy in the survival game can be represented as a probability distribution on i (or j) for each round, as a function of the past; this is the so-called "behavior strategy" form. We shall call a mixed strategy locally optimal if for every k the probabilities it prescribes for i_k (or j_k) are optimal in the matrix game $||v(r_{k-1} + a_{ij})||$, v being the value function of the survival game. Locally optimal strategies exist whenever the value function exists, but they need not be optimal, nor are optimal strategies necessarily locally optimal.

[Thus, in Example 4, consider the strategy that picks $i_k = 2$ if and only if $r_{k-1} > 1$. It is clearly locally optimal, since the value function is identically 1, but it is not optimal if $R > 2$ and $Q < 1$. Again, in Example 5, the strategy that always chooses $i_k = 3$ is locally optimal but not optimal if $Q < 1$ and $r_0 > 1$. (In this example the value depends on Q.)
In Example 6, the mixed strategy that prescribes the

$$\begin{pmatrix} 1 \\ -1 \end{pmatrix} \qquad \begin{pmatrix} 1 & -1 \\ -1 & 1 \\ 0 & 0 \end{pmatrix} \qquad \begin{pmatrix} 1 & -1 & 1 \\ -1 & 1 & 0 \\ -1 & -1 & 0 \end{pmatrix}$$

Example 4 Example 5 Example 6

probabilities $(1/3,\ 1/3,\ 1/3)$ for i_k if $j_{k-1} = 3$ and the probabilities $(1/2,\ 1/2,\ 0)$ if $j_{k-1} = 1$ or 2, or if $k = 1$, is optimal for Player I;

but it is not locally optimal, since it fails to take full advantage of the occasions when Player II makes the "mistake" of playing $j = 3$.]

A semimartingale may be defined as a sequence of random variables $\{x_k\}$ such that the conditional expectation of each term is greater than or equal to the preceding term, thus:

$$E\{x_k | x_{k-1}, \ldots, x_0\} \geq x_{k-1}.$$

A fundamental theorem ([5], page 324) implies that a bounded semi-martin-gale converges with probability 1, and that its limit x_∞ satisfies

$$E\{x_\infty | x_0\} \geq x_0.$$

For our purposes, "bounded" can be taken to mean that the x_k themselves are bounded, uniformly in k, although the results stated are valid under much weaker conditions.

Let ϕ be any bounded solution of (3). We define a local ϕ-strategy to be a mixed strategy that always prescribes optimal probabilities for the games $||\phi(r_{k-1} + a_{ij})||$. Thus, in this terminology, a locally optimal strategy is a local v-strategy. If Player I uses a local ϕ-strategy against an arbitrary strategy of Player II, then the sequence $\{\phi(r_k)\}$ that is generated is a bounded semimartingale. (Note that $E\{\phi(r_k)|r_{k-1}, \ldots, r_0\} \geq \phi(r_{k-1})$ implies $E\{\phi(r_k)|\phi(r_{k-1}), \ldots, \phi(r_0)\} \geq \phi(r_{k-1})$, even though ϕ may not be one-one.) Hence we have convergence with probability 1, and

$$E\left\{\lim_{k\to\infty} \phi(r_k)|r_0\right\} \geq \phi(r_0).$$

Now if ϕ satisfies (4) as well, the left side of this inequality can be expressed as

$$0 \cdot \text{prob}\{\text{I is ruined}\} + 1 \cdot \text{prob}\{\text{II is ruined}\} + \theta \cdot \text{prob}\{\text{both survive}\},$$

where θ is some number between 0 and 1. Hence:

$$\text{(5)} \qquad \text{prob}\{\text{II is ruined}\} \geq \phi(r_0) - \theta \cdot \text{prob}\{\text{both survive}\};$$

$$\text{(6)} \qquad \text{prob}\{\text{I survives}\} \geq \phi(r_0) + (1-\theta) \cdot \text{prob}\{\text{both survive}\}.$$

Thus such a strategy for Player I guarantees that he will survive with probability $\geq \phi(r_0)$. If we could show that double survival has probability zero, at least for some particular local ϕ-strategy of Player I, then it would follow that he can guarantee himself an expected payoff of $\phi(r_0)$, or more, regardless of the other's strategy, and regardless of Q. A similar argument for Player II would then establish the existence of a value and optimal strategies for the survival game, independent of Q.

In attempting to carry out a proof on the above lines, one might hope to start with an arbitrary local ϕ-strategy and

(i) use the known convergence of $\{\phi(r_k)\}$ to establish convergence of $\{r_k\}$; then

(ii) use the convergence of $\{r_k\}$ to show that the game must end; all with probability 1.

Unfortunately, neither (i) nor (ii) is unconditionally valid. In Section 3 we proceed by way of strictly monotonic approximants, for which (i) is valid, and obtain thereby the existence of the value. In Section 4 we obtain the existence of optimal strategies by working with a restricted class of "special ϕ-strategies", which make $\{r_k\}$ converge even when ϕ is not strictly monotonic. However, in both proofs it is necessary to assume that none of the a_{ij} is zero, in order to make convergence of $\{r_k\}$ equivalent to termination of play (step (ii)).

In Section 5 we drop the zero-free condition on $||a_{ij}||$, and find that a value still exists if Q is sufficiently regular. However, the value may depend on Q (see Examples 1, 2, 5 above), and the players may not have optimal strategies (Example 7 below). Our proof parallels the one in Section 3 (strictly monotonic approximants), but is based on a more complicated functional equation, to be discussed there.

Finally, in Section 6 we will derive some estimates for the value function that have much in common with the well-known approximate solutions of the classic "gambler's ruin" problem. They have simple analytic forms, in contrast to the sharply discontinuous nature of the exact value functions (see Examples 8 and 9 below). The estimates become more precise if R is made large compared to the a_{ij}, and they give exact information if the a_{ij} are all ±1, or ±1 and 0. They also provide strategies that are approximately optimal.

It should be noted that Sections 3, 4, 5, and 6 are essentially independent of one another.

[In Example 7, Player I can win with probability approaching 1 if he always chooses i_k according to the distribution $(1-\varepsilon, \varepsilon-\varepsilon^2, \varepsilon^2)$, with ε small but positive. However, if $Q<1$ he has no strictly optimal strategy. Example 8 illustrates in a simple way some of the possibilities for the value function $v(r)$. Under optimal play the first player's fortune descr:: es a random walk on $(0, R)$ with $+1$ and $-a$ having equal probability.

$$\begin{pmatrix} 1 & 0 & 0 \\ -1 & 1 & 0 \\ -1 & -1 & 1 \end{pmatrix} \qquad \begin{pmatrix} 1 & -a \\ -a & 1 \end{pmatrix} \qquad \begin{pmatrix} 1 & -a+\varepsilon \\ -a & 1 \end{pmatrix}$$

Example 7 — Example 8 — Example 9

The value is just the probability of absorption at R. If a is rational then the value is a finite step-function, which can be determined exactly by solving a certain system of linear equations. But if a is irrational (with $R > 1 + a > 1$), then the value function is discontinuous on a set of points everywhere dense in $(0, R)$; it is strictly monotonic; and its derivative is almost everywhere 0. In Example 9, let a be irrational, and assume $0 < \varepsilon < a < R - 1$. The value function is again strictly monotone, with discontinuities everywhere dense between 0 and R. We no longer have a simple random walk as above, and we no longer have a good description of the value function. Whether the derivative vanishes almost everywhere in this case is an open question.]

§2. SOLUTIONS OF THE FUNCTIONAL EQUATIONS

A monotonic solution to (3), (4) can be constructed by an iterative procedure. Define ϕ_0 by:

$$\phi_0(r) = \begin{cases} 0 & \text{if } r < R \\ 1 & \text{if } r \geq R \end{cases}$$

and let $\phi_n = T^n \phi_0$, where the transformation T is given by:

$$T\phi(r) = \begin{pmatrix} \text{val}||\phi(r + a_{ij})|| & 0 < r < R \\ \phi(r) & r \leq 0,\ r \geq R. \end{pmatrix} \tag{7}$$

It is clear that ϕ_n can be interpreted as the value function of the finite, truncated game in which Player I loses unless he succeeds in ruining his opponent in n moves or less.

LEMMA 1. The sequence $\{\phi_n\}$ just defined converges pointwise to a monotonic solution of (3), (4).

PROOF. By construction the fixed points of (7) are solutions of (3), and conversely. Since T is continuous, it suffices to show that $\lim \phi_n$ exists and is monotonic. This is accomplished by showing inductively that $\phi_n(r)$ is monotonic increasing in both n and r. The details present no difficulty whatever. (Compare the much harder proof of Lemma 5 below.)

Let v_0 denote the limit of the ϕ_n, and let v_1 denote the limit of the similar (descending) sequence $\{T^n \phi_0'\}$, beginning with the function

$$\phi_o'(r) = \begin{cases} 0 & \text{if } r \leq 0 \\ 1 & \text{if } r > 0. \end{cases}$$

THEOREM 1. If $Q \equiv 0$ then the value of the survival game exists and is equal to $v_0(r_0)$. If $Q \equiv 1$ then the value exists and is equal to $v_1(r_0)$.

PROOF. Player I can guarantee that

$$\text{prob}\{\text{II is ruined}\} \geq \phi_n(r_0)$$

by following an optimal strategy for the nth truncated game (and playing arbitrarily after the nth move). On the other hand, Player II can guarantee that

$$\text{prob}\{\text{II survives}\} \geq 1 - v_0(r_0)$$

by adopting a local v_0-strategy (see (6) above). But the payoff of the $Q \equiv 0$ game depends solely on whether Player II survives or not. Therefore $v_0(r_0)$ is its value. The other case is similar.

Note that the proof provides an optimal strategy for Player II, but not Player I, if $Q \equiv 0$. The existence of this optimal strategy, and of the value, could have been deduced from the lower semi-continuity of the payoff, as a function of the pure strategies (see [9]). A similar remark applies to the $Q \equiv 1$ game.

THEOREM 2. If (3), (4) have a unique solution ϕ, then the value of the survival game exists and is equal to $\phi(r_0)$, independently of Q.

PROOF. As before, Player I can ensure that

$$\text{prob}\{\text{II is ruined}\} \geq \phi_n(r_0).$$

Similarly Player II can ensure that

$$\text{prob}\{\text{I is ruined}\} \geq 1 - \phi'_n(r_0).$$

But $lim\, \phi_n = v_0 = \phi = v_1 = lim\, \phi'_n$; hence $\phi(r_0)$ is the value of the game.

Note that this time we do not obtain an optimal strategy for either player.

The next lemma identifies v_0 and v_1 as the "extreme" solutions of (3), (4); and incidentally establishes a converse to Theorem 2.

LEMMA 2. If ϕ is any solution of (3), (4) that is bounded between 0 and 1, then $v_0 \leq \phi \leq v_1$.

PROOF. We observe that $\phi_0 \leq \phi$, and that $\phi_n \leq \phi$ implies $T\phi_n \leq \phi$. Hence $v_0 \leq \phi$. Symmetrically, $v_1 \geq \phi$.

COROLLARY. If the value function of the survival game exists and is independent of Q, then it is the only solution of (3), (4) bounded between 0 and 1.

The last provision is necessary, since "spurious" unbounded solutions do sometimes occur.

The next lemma shows that v_0 and v_1 usually have jumps at 0 and R, and characterizes the exceptions in terms of the matrix $||a_{ij}||$.

LEMMA 3.

(A) The following are equivalent:

(i) $v_1(r)$ is continuous at $r = R$;
(ii) $v_1(r) \equiv 1$ for $0 < r < R$;
(iii) $||a_{ij}||$ has a nonnegative row.

(B) The following are equivalent:

(i) $v_0(r)$ is continuous at $r = R$;
(ii) $v_0(r) \equiv 1$ for $0 < r < R$;
(iii) every set of columns of $||a_{ij}||$, considered as a submatrix of $||a_{ij}||$, has a nonnegative row, not all zero.

Corresponding statements hold concerning continuity of v_0 and v_1 at $r = 0$.

PROOF. (A) Obviously (iii) $\Longrightarrow$ (ii) $\Longrightarrow$ (i). If (iii) is false there is a negative entry in each row. A strategy of playing all columns with equal probability, on every move, gives Player II a probability $\geq n^{[-R/a]}$ of winning, where n is the number of columns and a is the smallest non-zero $|a_{ij}|$. This gives a positive lower bound for $1 - v_1(r_0)$, independent of r_0, and makes v_1 discontinuous at R. Hence (i) $\Longrightarrow$ (iii).

(B) Obviously (ii) $\Longrightarrow$ (i). If (iii) is false there is a set of s columns on which player II can distribute his choices with equal probabilities $1/s$, giving him a probability $\geq s^{[-R/a]}$ of surviving. Hence $v_0(r_0)$ is bounded away from 1 and v_0 is discontinuous at R. Hence (i) $\Longrightarrow$ (iii). To complete the proof, suppose that (iii) holds but not (ii). Choose $r^* > 0$ so that $v_0(r^*) < v_0(r^* + a)$. Then $v_0(r^*)$ is strictly less than $v_0(r^* + a_{ij})$ whenever a_{ij} is positive. Let η be an optimal mixed strategy for II in the matrix game $||v_0(r^* + a_{ij})||$; let S be the set of columns j with $\eta_j > 0$; and let i_S be the nonnegative subrow, now all zero, whose existence is asserted by (iii). Then $v_0(r^*) \leq v_0(r^* + a_{i_s j})$ holds for j in S, with strict inequality at least once, and

$$\eta_j v_0(r^*) \leq \eta_j v_0(r^* + a_{i_s j})$$

holds for <u>all</u> j, with strict inequality at least once. Summing over j, and recalling the optimality of η, we obtain:

$$v_0(r^*) < \sum_j \eta_j v_0(r^* + a_{i_s j}) \leq \mathrm{val}||v_0(r^* + a_{ij})||.$$

But v_0 is a solution of (3), making the first and last terms equal. This contradiction establishes (iii) $\Longrightarrow$ (ii). (An alternative proof that (iii) implies (ii) could be obtained from the discussion below.)

COROLLARY. If

$$\max_i \min_j a_{ij} < 0 < \min_j \max_i a_{ij}$$

then every bounded solution of (3), (4) has jumps at 0 and R.

It may be of interest to describe some near-optimal strategies for Player I in the event that (i), (ii), (iii) of (B) hold. (Compare Example 7 above.) Let S_0 be the set of all columns; let i_0 be a row nonnegative and not identically zero on S_0; let S_1 be the subset of S_0 on which $a_{i_0 j} = 0$; and so on. Then we have $S_0 \supset \ldots \supset S_p \supset S_{p+1} = \phi$, for some p (proper inclusion all the way), and moreover the $i_0, \ldots, i_p$ are all distinct. Then it is easy to show that the probabilities:

$$\xi_{i_0} = 1 - \varepsilon, \ \xi_{i_1} = \varepsilon - \varepsilon^2, \ldots, \xi_{i_{p-1}} = \varepsilon^{p-1} - \varepsilon^p, \ \xi_{i_p} = \varepsilon^p,$$

all other $\xi_i = 0$, if used repeatedly by Player I, guarantee with probability $\geq 1 - \varepsilon$ that the first non-zero a_{ij} to occur will be positive. Hence Player I wins with probability $\geq (1-\varepsilon)^{-[-(R-r_0)/a]}$, no matter what Player II does. This bound goes to 1 as $\varepsilon \to 0$.

§3. EXISTENCE OF A VALUE WHEN $||a_{ij}||$ IS ZERO FREE

This section will be devoted to the proof of the following theorem;

THEOREM 3. If $||a_{ij}||$ is zero-free, then the value of the survival game exists and is independent of the payoff Q assigned to nonterminating play.

During the proof we shall work with certain generalized survival games, having more general, bounded payoff functions $P^*(r)$ in place of the $P(r)$ of (2). The other elements of the game, namely $||a_{ij}||, Q, R$ and r_0, remain as before. The functional equation (3) is still applicable, but with new boundary conditions:

$$(4^*) \qquad \phi(r) = P^*(r) \ \ r \leq 0, \ \ r \geq R.$$

LEMMA 4. Suppose that (3), (4*) have a strictly monotonic solution ϕ^*. Then, if $||a_{ij}||$ is zero-free, the value of the generalized survival game exists and is equal to $\phi^*(r_0)$.

PROOF. Let Player I use a local ϕ^*-strategy and Player II an arbitrary strategy. Then $\{\phi^*(r_k)\}$ is a bounded semimartingale, and converges with probability 1. Because ϕ^* is strictly monotonic, $\{r_k\}$ also converges. This means that play terminates, since, with none of the $a_{ij} = 0$, the limit of $\{r_k\}$ must be outside $(0, R)$. Hence

$$E\{P^*(lim\, r_k)\} = E\{\phi^*(lim\, r_k)\} \geq \phi^*(r_0).$$

A similar argument for Player II completes the proof.

We shall consider functions P^* of the following form only (until Section 6):

$$P^*(r) = \begin{cases} \varepsilon(r - R - A) & \text{if} \quad r \leq 0 \\ 1 + \varepsilon(r - R - A) & \text{if} \quad r \geq R, \end{cases}$$

where $A = max|a_{ij}|$, and ε is a positive constant. As $\varepsilon \to 0$ these functions approach $P(r)$ from below. Imitating the construction of Section 2, we define $\phi_n^* = T^n\phi_0^*$ where T is the same transformation (7), and ϕ_0^* is given by:

$$\phi_0^*(r) = \begin{cases} \varepsilon(r - R - A) & \text{if} \quad r < R \\ 1 + \varepsilon(r - R - A) & \text{if} \quad r \geq R. \end{cases}$$

If it exists, $lim\,\phi_n^*$ is obviously a solution of (3), (4*).

LEMMA 5. If $\text{val}||a_{ij}|| \geq 0$, and if v_0 is not continuous at R, then for sufficiently small ε the sequence $\{\phi_n^*\}$ just defined converges pointwise to a strictly monotonic solution of (3), (4*).

PROOF. To show that the limit exists we observe first that

$$\begin{aligned} \phi_1^*(r) &= \text{val}||\phi_0^*(r + a_{ij})|| \geq \text{val}||\varepsilon(r - R - A + a_{ij})|| \\ &= \varepsilon(r - R - A) + \varepsilon \cdot \text{val}||a_{ij}|| \geq \varepsilon(r - R - A) \\ &= \phi_0^*(r) \end{aligned}$$

if $0 < r < R$, and $\phi_1^*(r) = \phi_0^*(r)$ otherwise. Moreover, $\phi_n^* \geq \phi_{n-1}^*$ implies $\phi_{n+1}^* = T\phi_n^* \geq T\phi_{n-1}^* = \phi_n^*$. Since the sequence is obviously bounded, it therefore converges. To show that the limit is strictly monotonic we shall prove inductively that for ε sufficiently small the function $\phi_n^*(r) - \varepsilon r$ is monotonic in r for each n. This is trivial for ϕ_0^*; assume it for ϕ_{n-1}^*.

CASE 1. Take $0 < r < s < R$. Then we have:

$$\begin{aligned} \phi_n^*(r) - \varepsilon r &= \text{val}||\phi_{n-1}^*(r + a_{ij}) - \varepsilon r|| \\ &\leq \text{val}||\phi_{n-1}^*(s + a_{ij}) - \varepsilon s|| \\ &= \phi_n^*(s) - \varepsilon s. \end{aligned}$$

CASE 2. Take $r \leq 0 < s < R$. Then we have:

$$\begin{aligned} \phi_n^*(r) - \varepsilon r &= \phi_{n-1}^*(r) - \varepsilon r \\ &\leq \phi_{n-1}^*(s) - \varepsilon s \\ &\leq \phi_n^*(s) - \varepsilon s \end{aligned}$$

(using the first part of this proof in the last step).

CASE 3. Take $0 < r < R \leq s$. Note that $\phi_0^* \leq v_0$ over the entire range of interest $(-A, R + A)$, and hence, by induction, that $\phi_n^* \leq v_0$. Since the latter is assumed discontinuous at R, we can select ε so that $\varepsilon A \leq 1 - v_0(R-)$. Then we have:

$$\phi_n^*(r) - \varepsilon r \quad \leq \quad \phi_n^*(R-) - \varepsilon R$$

$$\begin{aligned} &\leq v_0(R-) - \varepsilon R \\ &\leq 1 - \varepsilon A - \varepsilon R \\ &= \phi_n^*(s) - \varepsilon s \end{aligned}$$

(using Case 1 in the first step).

The other cases, namely $r < s \leq 0$, $r \leq 0 < R \leq s$, and $R \leq r < s$, are trivial. This completes the proof of Lemma 5.

PROOF OF THEOREM 3. There is no loss of generality in assuming that $\text{val}||a_{ij}|| \geq 0$. The theorem is trivial if $v_0 \equiv 1$ in $(0, R)$; we may therefore assume (Lemma 3) that v_0 is discontinuous at R. Lemmas 4 and 5 now give us well-defined value functions ϕ^* for our class of generalized games, for ε sufficiently small. As $\varepsilon \to 0$ their payoffs P^* converge uniformly to P, our original payoff function. It follows that

$$\lim_{\varepsilon \to 0} \phi^*(r_0)$$

exists and is the value of the original game. This number is obviously independent of Q. This completes the proof.

As yet we know nothing about the existence of optimal strategies, unless $Q \equiv 0$ or $Q \equiv 1$. As Example 4 showed, local v-strategies need not be optimal. However, the local ϕ^*-strategies, for given ε, are nearly optimal: it can be shown that they guarantee an expected payoff within $\varepsilon(R + 2A)$ of the true value.

COROLLARY. If $||a_{ij}||$ is zero-free then (3), (4) have a unique solution bounded between 0 and 1.

PROOF. Theorem 3 and the Corollary to Lemma 2.

§4. EXISTENCE OF OPTIMAL STRATEGIES

In this section we shall find strategies for both players, under the assumption that $||a_{ij}||$ is zero-free. The strategies are locally optimal, with the added property that they force play to terminate with probability one. Their optimality is therefore independent of Q.

The strategies in question are constructed as follows. Let ϕ be a monotonic solution of (3), (4). For each number c for which the set $C = \phi^{-1}(c)$ is not empty, let $\xi(c)$ be an optimal strategy of Player I in the matrix game

(8) $$||\inf_{s \in C} \phi(s + a_{ij})||.$$

Define a mixed strategy for the survival game by the rule: choose i_k according to the probability distribution $\xi(\phi(r_{k-1}))$. Such a strategy will be called a special ϕ-strategy of Player I. Note that the same probabilities must be used each time the same value of $\phi(r)$ comes up.

Special ϕ-strategies of Player II are defined similarly with "sup" instead of "inf". We shall see presently that a special ϕ-strategy is also a local ϕ-strategy. First we state our main result.

THEOREM 4. If ϕ is any monotonic solution of (3), (4), and if $||a_{ij}||$ is zero-free, then the special ϕ-strategies are optimal, independently of Q, and the value of the game is $\phi(r_0)$.

As stated, Theorem 4 is independent of and includes Theorem 3, and this independence will be maintained throughout the proof, which takes up the rest of this section. Consequently we have a separate proof of the existence of a value in the zero-free case. Of course, ϕ is actually unique and equal to v, and our real object is only to show that the special v-strategies are optimal.

LEMMA 6. A special ϕ-strategy is also a local ϕ-strategy.

PROOF. We must show that $\xi(\phi(r))$ is optimal in $||\phi(r + a_{ij})||$, for all r in $(0, R)$. But $||\phi(r + a_{ij})||$ majorizes (8), and both matrices have the same value $\phi(r)$, by (3) and the fact that ϕ was assumed monotonic. Since $\xi(\phi(r))$ is optimal for (8) by definition, it is also optimal for the matrix $||\phi(r + a_{ij})||$.

The proof of Theorem 4 will be based on the following concept. Given a special ϕ-strategy $\xi(c)$ for Player I define the "punishment function":

$$\pi(r, j) = \sum_i \xi_i(c)\phi(r + a_{ij}) - c,$$

where $c = \phi(r)$. By Lemma 6, π is always non-negative. If Player II follows a local ϕ-strategy then π will always be zero. (Thus π measures the expected amount Player II will be punished for choosing the j-th column.)

LEMMA 7. If Player I uses a special ϕ-strategy against any strategy of Player II then every possible play of the survival game has the property that, for each $k > 0$, one of the following is true:

(a) $r_{k-1} \leq 0$ or $r_{k-1} \geq R$;

(b) $\phi(r_{k-1}) = 0$;

(c) $\pi(r_{k-1}, j_k) > 0$;

(d) $r_k \geq r_{k-1} + a$, where $a = \min_{i,j} |a_{ij}|$;

(e) $\phi(r_k) \neq \phi(r_{k-1})$.

The importance of this lemma lies in the fact that it sharply restricts the possibility of a non-terminating play in which $\{\phi(r_k)\}$ converges.

PROOF. We will assume that (a), (b), (d), and (e) are false and prove that (c) is true. Since (a) is false we have

$$r_k = r_{k-1} + a_{i_k j_k}.$$

Since (d) is false this implies that

$$r_k < r_{k-1} \text{ or } a_{i_k j_k} < 0.$$

Let $c = \phi(r_{k-1})$, $C = \phi^{-1}(c)$. Since (b) is false it follows that this set C has a lower bound, and therefore that

$$\inf_{s \in C} \phi(s + a_{i_k j_k}) < c.$$

Since (e) is false we have

$$c = \phi(r_k) = \phi(r_{k-1} + a_{i_k j_k}).$$

Therefore, in the inequality

$$\inf_{s \in C} \phi(s + a_{i j_k}) \leq \phi(r_{k-1} + a_{i j_k}),$$

the strict inequality holds for $i = i_k$. The fact that the row i_k is actually played by Player I on the k-th move implies that its probability $\xi_{i_k}(c)$ is positive. Therefore

$$\sum_i \xi_i(c) \inf_{s \in C} \phi(s + a_{i j_k}) < \sum_i \xi_i(c) \phi(r_{k-1} + a_{i j_k}).$$

But since $\xi(c)$ is an optimal strategy for the matrix game

$$\| \inf_{s \in C} \phi(s + a_{ij}) \|$$

with value c, the left hand side of this inequality is $\geq c$. Therefore

$$\pi(r_{k-1}, j_k) = \sum_i \xi_i(c) \phi(r_{k-1} + a_{i j_k}) - c$$

is positive. This proves assertion (c) and completes the proof of Lemma 7.

LEMMA 8. If Player I uses a special ϕ-strategy against any strategy of Player II, then with probability one the sequence $\{\pi_k\} = \{\pi(r_{k-1}, j_k)\}$ converges to 0.

PROOF. (Compare [5], page 297, Theorem 1.2 (i).) Consider the sum of the π_k. For each $n \geq 1$ we have:

$$\begin{aligned} E\left\{\sum_1^n \pi_k\right\} &= E\left\{\sum_1^n [E\{\phi(r_k)|r_{k-1}\} - \phi(r_{k-1})]\right\} \\ &= \sum_1^n [E\{\phi(r_{k-1})\} - E\{\phi(r_{k-1})\}] \\ &= E\{\phi(r_n)\} - E\{\phi(r_0)\} \\ &\leq 1. \end{aligned}$$

It follows that the probability of the infinite sum exceeding any given bound M is $\leq 1/M$. Hence, with probability one the series has a finite sum and $\{\pi_k\}$ converges to 0.

PROOF OF THEOREM 4. The theorem is easy if there is a positive row or negative column in $||a_{ij}||$. We therefore assume

$$\max_i \min_j a_{ij} < 0 < \min_j \max_i a_{ij}.$$

Let player I adopt a special ϕ-strategy, and Player II an arbitrary strategy. The play of the game that then occurs can be described by the pair of sequences $\{i_k\}$, $\{j_k\}$, which we shall regard as the underlying random variable. They are sufficient to determine three other important sequences $\{r_k\}$, $\{\pi_k\}$, and $\{\phi(r_k)\}$. The last-named is a bounded semimartingale, by Lemma 6, and therefore the set of plays $\{i_k\}$, $\{j_k\}$ for which it fails to converge has probability zero. The set of plays for which $\{\pi_k\}$ fails to converge to 0 also has probability zero, by Lemma 8. We shall prove that every play outside these two sets terminates. By (5) this will imply that Player I's special ϕ-strategy assures him an expected payoff $\geq \phi(r_0)$.
The corresponding argument for the other player will complete the proof.

Consider therefore a play $\{i_k\}$, $\{j_k\}$ in which both

$$\{\phi(r_k)\} \to c \tag{9}$$

and

$$\{\pi_k\} \to 0. \tag{10}$$

If $c = 0$ or 1 then play must terminate, because of the fact that ϕ is discontinuous at 0 and R (see the Corollary of Lemma 3, Section 2). Thus our object will be to show that the hypothesis $0 < c < 1$ leads to a contradiction. Let $\{k_n\}$ be the sequence of indices k such that $\phi(r_k) \neq c$ or $\pi_k \neq 0$, or both, and let $s_n = r_{k_n}$. Lemma 7 shows that $\{k_n\}$ and $\{s_n\}$ have infinitely many terms, since alternatives "(a)" and "(b)" are excluded by hypothesis, while any unduly long chain of "(d)" will take

r_k out of the interval $C = \phi^{-1}(c)$. Thus, instances of "(c)" or "(e)" must occur regularly, giving us $\pi_k \neq 0$, or $\phi(r_k) \neq \phi(r_{k-1})$. In fact, consecutive terms of $\{k_n\}$ can not differ by more than $[(\gamma/a)+2]$, where γ is the length of C, $a = min\,|a_{ij}|$, and $[x]$ is the greatest integer $\leq x$.

Let us now examine the possible limit points of the sequence $\{s_n\}$, bearing in mind properties (9) and (10). First we have the upper and lower endpoints of C; call them u and ℓ respectively. (If C is empty, a short argument shows that (9) is contradicted.) Secondly, there are the points within C, in the neighborhood of which π_k can be arbitrarily small, but positive. There are only a finite number of such points, since, for each j, $\pi(r,j)$ is a monotonic increasing function of r in C; denote these points by $y_3, y_4, \ldots, y_p$. Let δ be a small positive constant (it has an exact value, which will be defined later), and let $Y_1, Y_2, \ldots, Y_p$ be a collection of intervals, variously open and closed, but all of length δ, defined as follows:

$$Y_1 = \begin{cases} [u, u+\delta) & \text{if} \quad u \notin C \\ (u, u+\delta) & \text{if} \quad u \in C \end{cases}$$

$$Y_2 = \begin{cases} (\ell-\delta, \ell] & \text{if} \quad \ell \notin C \\ (\ell-\delta, \ell) & \text{if} \quad \ell \in C \end{cases}$$

$$Y_i = (y_i, y_i+\delta] \quad 3 \leq i \leq p.$$

Let Y denote their set-theoretic union. By (9), (10), there is an n_0 (depending on δ) such that $s_n \in Y$ for all $n > n_0$.

Now define δ to be the smallest nonzero number of the form:

$$\left|\sum_{\nu=1}^{N} a_{i_\nu j_\nu}\right|, \qquad N \leq p[(\gamma/a)+2]. \tag{11}$$

The effect of this definition is to ensure that when $|m-n| \leq p$, the difference $|s_m - s_n|$ is either 0 or $\geq \delta$. It follows that, <u>for</u> $n \geq n_0$, <u>all</u> s_n <u>lying in a given</u> Y_i <u>are equal</u> (proof below). Hence some s^* appears infinitely often in the sequence $\{s_n\}$. If $s^* \notin C$ then (9) is contradicted. If $s^* \in C$ then (10) is contradicted, since π_{k_n} is infinitely often positive, and bounded away from 0 by the smallest nonzero $\pi(s^*, j)$. This is the desired contradiction.

Finally, we have to prove the statement underlined above. Call a pair (s_m, s_n) <u>conflicting</u> if they lie in the same Y_i but are unequal. Either the statement in question is true or there is a conflicting pair (s_m, s_n), coming after s_{n_0}, and spanning no other conflicting pair $(s_{m'}, s_{n'})$, $m \leq m' < n' \leq n$. Let $m_0 = m$ and let m_{j+1} be the last integer $\leq n$ such that $s_{m_{j+1}}$ and s_{m_j+1} lie in a common interval Y_i. Then the ascending sequence

$$m_0, m_1, m_2, \ldots, m_{q-1}, m_q = n$$

has at most $p+1$ terms, and has the property that $s_{m_{j+1}} = s_{m_j+1}$ for each j. We see that each s_{m_j} is the sum of $s_{m_{j-1}}$ and at most $[(\gamma/a)+2]$ increments $a_{i_\nu j_\nu}$. Hence $|s_m - s_n|$ is of the form (11), and is either 0 or $\geq \delta$. This contradicts the assumption that (s_m, s_n) was a conflicting pair.

§5. EXISTENCE OF A VALUE IN GENERAL

When there are zeros in the $||a_{ij}||$ matrix a unique solution to the fundamental equations (3), (4) is no longer assured, and the value of the survival game may depend on the double-survival payoff Q. In this section we show that the value does exist if Q is sufficiently regular. Our proof makes use of a new pair of functional equations involving the notion of recursive game (equations $(\bar{3})$, $(\bar{4})$ below), which reduce to (3), (4) if $||a_{ij}||$ is zero-free, and which always have a unique solution.

The restriction on Q consists in assuming that whenever $\{r_k\}$ converges to a limit r within $(0, R)$, the payoff depends only on that limit: $Q = \bar{P}(r)$. (Of course $\{r_k\}$ can converge to r only if the equality $r_k = r$ holds for all sufficiently large k.) We assume $\bar{P}$ to be monotonic increasing and, of course, bounded between 0 and 1. It is convenient to amalgamate it with the old function P, which was defined only outside $(0, R)$, and denote both by $\bar{P}$. The still-arbitrary portion of Q will be denoted by $\bar{Q}$. Thus, the principal distinction in the present set-up is between convergent play (payoff $\bar{P}$) and nonconvergent play (payoff $\bar{Q}$), replacing the former distinction between terminating play (payoff P) and nonterminating play (payoff Q).

Note that the present arrangement still includes the important special cases $Q \equiv$ constant. On the other hand, the mischievous Gale-Stewart functions (Example 3 in Section 1) are kept out.

THEOREM 5. If the convergent-play payoff is a monotonic function $\bar{P}$ of the limit of the first player's fortune, then the value of the survival game exists and is independent of the payoff $\bar{Q}$ for nonconvergent play.

The proof follows the same general lines as the proof of Theorem 3 in Section 3. However we must begin with some new definitions and notation.

By an elementary recursive game $[M, p]$ we shall mean the following: A matrix $||m_{ij}|| = M$ is given, each entry being either a number or the symbol Ⓡ. Players I and II choose i_1 and j_1 respectively. If $m_{i_1 j_1}$ is a number the game ends, and II pays I the indicated amount. If $m_{i_1 j_1}$ is Ⓡ there is no payment, and the players go back to the beginning and make new choices i_2 and j_2, etc. In case of infinite repetition the payoff is the number p (a constant).

The above is a special case of the recursive games defined by Everett [6], except for the "p" feature. (Everett effectively assumes that $p = 0$. However, $[M, p]$ is obviously equivalent to $[M', 0]$, with M' obtained by subtracting p from each numerical entry of M.) Translating the results of [6] we find that the elementary recursive game has a value, which we shall denote by $\overline{\text{val}}[M, p]$, though perhaps not

optimal strategies.
The value satisfies the relation

$$x = \text{val}||m_{ij} : x||, \tag{12}$$

where $||m_{ij} : x||$ is the matrix obtained by inserting the numerical variable x in play of ® in $||m_{ij}||$. The solutions of (12) form a closed interval, and it develops that $\overline{\text{val}}[M, p]$ is the solution of (12) that is closest to p.

The new functional equations can now be formulated; they are:

$$\phi(r) = \overline{\text{val}}[M(\phi, r),\ \bar{P}(r)], \quad 0 < r < R \tag{$\bar{3}$}$$

and

$$\phi(r) = \bar{P}(r), \quad r \leq 0,\ r \geq R, \tag{$\bar{4}$}$$

where $M(\phi, R)$ is the matrix with entries:

$$m_{ij}(\phi, r) = \begin{cases} \phi(r + a_{ij}) & \text{if} \quad a_{ij} \neq 0 \\ \text{®} & \text{if} \quad a_{ij} = 0. \end{cases}$$

($\bar{4}$) and (4) are actually the same. If $||a_{ij}||$ is zero-free, then ($\bar{3}$) reduces to (3). In general, if ϕ is a solution of ($\bar{3}$), then (12) gives us:

$$\phi(r) = \text{val}||m_{ij}(\phi, r) : \phi(r)|| = \text{val}||\phi(r + a_{ij})||;$$

that is, ϕ is a solution of (3) as well. The converse is not true, however, since in fact ($\bar{3}$), ($\bar{4}$) always have a unique solution, while (3), (4) do not.

As in Section 2 (Lemma 1), we can construct a monotonic solution $\bar{v}_0$ to ($\bar{3}$), ($\bar{4}$) by iterating the transformation $\bar{T}$:

$$\bar{T}\phi(r) = \begin{cases} \overline{\text{val}}[M(\phi, r), \bar{P}(r)] & 0 < r < R \\ \phi(r) & r \leq 0,\ r \geq R, \end{cases} \tag{$\bar{7}$}$$

applied to the same initial function $\phi_0(r)$ which is 0 for $r < R$ and 1 for $r \geq R$. It is easily shown that the functions $\bar{T}^n\phi_0(r)$ are monotonic in r and form a bounded increasing sequence; the limit is the desired function $\bar{v}_0$. (It can be interpreted as the value function of the $\bar{Q} \equiv 0$ game (compare Theorem 1), but there is no point in establishing this fact now, in view of the stronger result which will be proved as Theorem 7.)

As before, we introduce certain generalized payoff functions $\bar{P}^*(r)$; they will be assumed monotonic increasing in $(-A, R + A)$. The symbols ($\bar{3}^*$), ($\bar{4}^*$), ($\bar{7}^*$) will refer to equations ($\bar{3}$), ($\bar{4}$), ($\bar{7}$) with $\bar{P}$ replaced by $\bar{P}^*$. The next lemma corresponds to Lemma 4 in Section 3.

LEMMA 9. Suppose that $(\bar{3}^*)$, $(\bar{4}^*)$ have a strictly monotonic solution $\bar{\phi}^*$. Then the value of the generalized survival game exists and is equal to $\bar{\phi}^*(r_0)$.

PROOF. Relative to a particular play of the generalized survival game define $k_0 = 0$ and let k_{n+1} be the first k (if any) such that $r_k \neq r_{k_n}$. The subsequence $\{s_n\} = \{r_{k_n}\}$ is finite in length if and only if $\{r_k\}$ converges. We now describe a "local recursive ε-optimal $\bar{\phi}^*$-strategy" for Player I; it resembles our previous "local" strategies, but is based on elementary recursive games instead of matrix games. Choose a sequence of positive numbers $\varepsilon_0, \varepsilon_1, \varepsilon_2, \ldots$ with sum ε. Let Player I begin by playing an ε_0-optimal strategy of the elementary recursive game $[M(\bar{\phi}^*, r_0),\ \bar{P}^*(r_0)]$. If and when that strategy runs out (after k_1 moves, in fact), let him continue with an ε_1-optimal strategy of $[M(\bar{\phi}^*, s_1),\ \bar{P}^*(s_1)]$, and so on. In general. on his $(k_n + 1)^{\text{th}}$ move, he will be commencing an ε_n-optimal strategy of $[M(\bar{\phi}^*, s_n),\ \bar{P}^*(s_n)]$. We wish to show that such a strategy, played against an arbitrary strategy Player II, causes $\{r_k\}$ to converge with probability one.

Define the <u>infinite</u> sequence $\{x_n\}$ as follows:

$$x_n = \begin{cases} \bar{\phi}^*(s_n) & \text{if } \{s_i\} \text{ is defined through } i = n, \\ \bar{P}^*(s_{n_0}) & \text{if } \{s_i\} \text{ stops at } i = n_0 < n. \end{cases}$$

Our construction ensures that, for $n = 1, 2, \ldots$,

$$E\{x_n | x_{n-1}, \ldots, x_0\} \geq x_{n-1} - \varepsilon_{n-1}.$$

Therefore the sequence $x_0, x_1 + \varepsilon_0, x_2 + \varepsilon_0 + \varepsilon_1, x_3 + \varepsilon_0 + \varepsilon_1 + \varepsilon_2$, etc. is a bounded semimartingale. We conclude that $\{x_n\}$ converges with probability one, with

$$E\{x_\infty | x_0\} \geq x_0 - \varepsilon.$$

However, $\{x_n\}$ can not converge if $\{s_i\}$ does not stop at some s_{n_0}, since the s_i oscillate and $\bar{\phi}^*$ is strictly monotonic. Hence $x_\infty = \bar{P}^*(s_{n_0}) = \bar{P}^*(\lim r_k)$, and we have:

$$E\{\bar{P}^*(\lim r_k)\} \geq \bar{\phi}^*(r_0) - \varepsilon.$$

The rest of the proof is obvious.

We now particularize $\bar{P}^*(r)$ to be $\bar{P}(r) + \varepsilon(r - R - A)$, where ε is a positive constant and $A = \max a_{ij}$. (Compare Section 3.) Using the same initial function as before:

$$\bar{\phi}_0^*(r) = \phi_0^*(r) = \begin{cases} \varepsilon(r - R - A) & \text{if } r < R \\ 1 + \varepsilon(r - R - A) & \text{if } r \geq R, \end{cases}$$

we generate a sequence $\{\bar{\phi}_n^*\} = \{\bar{T}^{*^n} \phi_0^*\}$ by iterating the new transformation $\bar{T}^*$, given by $(\bar{7}^*)$. The next lemma corresponds to Lemma 5.

LEMMA 10. If val$||a_{ij}|| \geq 0$, and if $\bar{v}_0$ is not continuous at R, then for sufficiently small ε the sequence $\{\bar{\phi}^*_n\}$ just defined converges to a strictly monotonic solution of $(\bar{3}^*, (\bar{4}^*)$.

The proof is essentially the same as the proof of Lemma 5. The substitution of elementary recursive games for matrix games causes trouble only at one spot: the proof of $\bar{\phi}^*_1 \geq \bar{\phi}^*_0$. The difficulty is resolved by an appeal to the following fact: $p > 0$ and val$||m_{ij} : 0|| > 0$ together imply that val$[M, p] \geq 0$.

LEMMA 11. If both v_0 and $\bar{P}$ have jumps at R, then so does $\bar{v}_0$.

PROOF. We use the fact that $\bar{v}_0$ is a solution of $(\bar{3}), (\bar{4})$. As in the proof of Lemma 11 we can find a "local recursive ε-optimal $\bar{v}_0$-strategy" for Player I that ensures that the sequence $\{x_n\}$ (as defined there, but with $\bar{P}$ for $\bar{P}^*$ and $\bar{v}_0$ for $\bar{\phi}^*$) converges with probability 1, and that

$$E\{x_\infty | x_0\} \geq \bar{v}_0(r_0) - \varepsilon.$$

This holds for any strategy of Player II; we shall consider a particular one. By Lemma 3 the jump in v_0 means that $||a_{ij}||$ has a set of columns that meets each row in subrow that contains a negative element, or is all zero. The same strategy for Player II used in the proof of Lemma 3, part B, guarantees a probability $\geq s[-R/a] = \delta > 0$ that $\{r_k\}$ will never increase. This means that with probability $\geq \delta$, $\{r_k\}$ will converge to a limit $r \leq r_0 < R$. Hence

$$E\{x_\infty | x_0\} \leq (1 - \delta) + \delta \cdot \bar{P}(R-).$$

This bound is < 1 because of the jump in $\bar{P}$ at R, and is independent of r_0. Thus, letting $\varepsilon \to 0$ we find that $\bar{v}_0(r_0)$ is bounded away from 1 for $0 < r_0 < R$, as was to be shown.

PROOF OF THEOREM 5. There is no loss of generality in assuming val$||a_{ij}|| \geq 0$. The theorem is trivial if v_0 is continuous at R, since then $v_0 = v_1$ by Lemmas 2 and 3; so we can assume that v_0 has a jump at R. Assume for the moment that $\bar{P}$ also has a jump at R. Then, applying Lemmas 13, 12 and 11 in that order, we find that the $\bar{P}^*$ games all have values as $\varepsilon \to 0$. The uniform convergence ensures that the original game also has a value, and it is clear that this value, being the limit of the $\bar{P}^*$ values, is independent of $\bar{Q}$. On the other hand, if $\bar{P}$ is continuous at R, then we can approximate it uniformly by a sequence of discontinuous, monotonic functions. The preceding argument applies to the latter, and passing to the limit completes the proof.

COROLLARY. Equations $(\bar{3}), (\bar{4})$ have a unique solution, assuming only that $\bar{P}$ is monotonic and satisfies (4).

PROOF. Let ϕ be any solution of $(\bar{3}), (\bar{4})$ and consider the game determined by $\bar{P}, \bar{Q}$, with $\bar{Q} \equiv 1$. A "local recursive ε-optimal ϕ-strategy" for Player I (see proof of Lemma 11) will guarantee him an expected payoff of at least $\phi(r_0) - \varepsilon$ in this game. Thus:

$$v(r_0) \geq \phi(r_0),$$

if v is its value function. But v is also the value function of the game defined by $\bar{Q} \equiv 0$, by Theorem 5. By symmetry we have:

$$v(r_0) \leq \phi(r_0).$$

Thus ϕ is uniquely determined.

§6. APPROXIMATIONS AND BOUNDS FOR THE VALUE FUNCTION

In this section we extend to games of survival some of the known results for random walks with absorbing barriers -- i.e., the gambler's ruin problem (see [7], Chapter 14). The random walk on $(0, R)$ with each step determined by the fixed random variable ξ leads naturally to a functional equation, highly reminiscent of our fundamental equation (3):

(13) $$\phi(r) = E\{\phi(r+\xi)\}, \quad 0 < r < R.$$

It is satisfied by several functions associated with the random walk; among them is the probability $p_R(r)$ that a particle starting at r will reach R before it reaches 0. This "absorption" probability is uniquely determined by (13) and the familiar boundary condition:

(4) $$\phi(r) = P(r), \quad r \leq 0,\ r \geq R;$$

assuming that ξ is not identically 0.

If it happens that $E\{\xi\} = 0$, then (13) has among its solutions all linear functions $A + Br$. Applying the two conditions $p_R(0) = 0$ and $p_R(R) = 1$ we get $A = 0$ and $B = 1/R$, or

$$p_R(r) \approx \frac{r}{R}, \quad 0 < r < R.$$

This is not exact because the particle will in general be absorbed beyond, not at, the barriers 0 and R. Taking this fact into account, we obtain rigorous estimates:

$$\frac{r}{R+\nu} \leq p_R(r) \leq \frac{r+\mu}{R+\mu}, \quad 0 < r < R,$$

where μ and ν are such that always $-\mu \leq \xi \leq \nu$.

If on the other hand $E\{\xi\} \neq 0$, then there will be a unique, nonzero λ_0 such that

(14) $$E\left\{e^{\lambda_0 \xi}\right\} = 1,$$

provided that ξ takes on both positive and negative values with positive probability (see [7], page 302, or [15], page 284). Then (13) has among its solution all functions of the form $A + Be^{\lambda_0 r}$. As before, this leads to an approximation:

$$p_R(r) \approx \frac{e^{\lambda_0 r} - 1}{e^{\lambda_0 R} - 1} \qquad 0 < r < R$$

and bounds:

$$\frac{e^{\lambda_0 r} - 1}{e^{\lambda_0 (R+\nu)} - 1} \leq p_R(r) \leq \frac{e^{\lambda_0 (r+\mu)} - 1}{e^{\lambda_0 (R+\mu)} - 1}, \qquad 0 < r < R.$$

The linear case first discussed (with $E\{\xi\} = 0$) corresponds to $\lambda_0 = 0$. Actually it is not an exceptional case; this becomes evident if we introduce the function f:

$$f(\lambda, x) = \begin{cases} (e^{\lambda x} - 1)/\lambda & \text{if} \quad \lambda \neq 0 \\ x & \text{if} \quad \lambda = 0, \end{cases}$$

which is continuous and monotonic increasing in λ for each x. (In fact for $x \neq 0$ the function $f(\lambda, x)$ is strictly monotonic in λ.) The relation (14) defining λ_0 becomes $E\{f(\lambda_0, \xi)\} = 0$, which has a unique solution in all cases. Write $F(x)$ for $f(\lambda_0, x)$. Then the approximation and bounds for p_R are simply $F(r)/F(R)$, $F(r)/F(R+\nu)$ and $F(r+\mu)/F(R+\mu)$, respectively, regardless of whether λ_0 is positive, negative, or zero. For games of survival we have some very similar results:

LEMMA 12. If $||a_{ij}||$ is zero-free and if

$$\max_i \min_j a_{ij} < 0 < \min_j \max_j a_{ij},$$

then there is a unique number λ_0 such that

$$\text{(15)} \qquad \text{val}||f(\lambda_0, a_{ij}|| = 0.$$

Moreover λ_0 and $\text{val}||a_{ij}||$ have opposite signs, or are both zero.

PROOF. The value of $||f(\lambda, a_{ij})||$ is a continuous and strictly monotonic function of λ, since none of the a_{ij} is 0. It tends to the limit $+\infty$ as $\lambda \to +\infty$ because of the positive element in each column, and to the limit $-\infty$ as $\lambda \to -\infty$ because of the negative element in each row. Therefore it has precisely one zero. The last part of the lemma follows from the fact that $\text{val}||f(0, a_{ij})|| = \text{val}||a_{ij}||$.

Again write $F(x)$ for $f(\lambda_0, x)$.

LEMMA 13. F is a solution of (3).

PROOF. Using the identity (valid for all λ):

$$f(\lambda, x+y) = f(\lambda, x) + e^{\lambda x} f(\lambda, y).$$

we have:

$$\begin{aligned} \mathrm{val}||F(r+a_{ij})|| &= F(r) + e^{\lambda_0 r}\mathrm{val}||F(a_{ij})|| \\ &= F(r) \end{aligned}$$

as required. The corollary which follows is proved by the same device.

COROLLARY. The local F-strategies are precisely those mixed strategies that use only probability distributions that are optimal in the matrix game $||F(a_{ij})||$.

THEOREM 6. If $||a_{ij}||$ is zero-free and if

$$\max_i \min_j a_{ij} < 0 < \min_j \max_i a_{ij}$$

then the value of the survival game is approximately equal to $F(r_0)/F(R)$. More precisely, we have:

(16) $$\frac{F(r_0)}{F(R+\nu)} \leq v(r_0) \leq \frac{F(r_0+\mu)}{F(R+\mu)},$$

where $\nu = \max a_{ij}$, $\mu = -\min a_{ij}$. The local F-strategies are approximately optimal, in the sense that Player I can enforce the lower bound of (16), and Player II the upper bound, by using them.

We remark that an all-positive row ($\max\min a_{ij} > 0$) or an all- negative column ($\min\max a_{ij} < 0$) trivializes the game. (These cases correspond to $\lambda_0 = -\infty$ and $+\infty$ respectively.)

PROOF OF THEOREM 6. Denote the indicated lower bound in (16) by $g(r_0)$. Clearly g is a strictly monotonic solution of (3), and the local F-strategies are also local g-strategies. If we set $P^*(r) = g(r)$ for r outside $(0, R)$ we have a "generalized survival game" in the sense of Section 3. By Lemma 4, $g(r_0)$ is its value, and the local F-strategies are optimal. But $P^* \leq P$ throughout the relevant intervals $(-\mu, 0]$ and $[R, R+\nu)$; therefore $g(r_0) \leq v(r_0)$, and the local F-strategies enforce at least the lower amount for Player I. The other bound is established in the same way.

The bounds (16) can sometimes be improved by exploiting special properties of the matrix. For example, inadmissible rows or columns of $||F(a_{ij})||$ can be disregarded in calculating μ and ν. Two other such results are the following:

COROLLARY 1. If r_0, R, and the a_{ij} are all integers, then μ and ν in (16) may be replaced by $\mu - 1$ and $\nu - 1$ respectively.

COROLLARY 2. If r_0 and R are integers and the a_{ij} are all ± 1, then the value of the game is exactly $F(r_0)/F(R)$, and the local F-strategies are optimal.

An equally exact result holds for arbitrary r_0 and R; it has the form $v(r_0) = F(r_0 + \bar{\mu})/F(R + \bar{\mu} + \bar{\nu})$, where $-\bar{\mu}$ and $R + \bar{\nu}$ are the unique absorption points of the process. However, $\bar{\mu}$ and $\bar{\nu}$ depend on r_0 in such a way that v is actually a step function, despite the continuity of F.

Two simple asymptotic results are of interest:

COROLLARY 3. If $R \to \infty$ with r_0 held fixed, the value of the game tends to a limit ℓ that is $= 0$ or satisfies

$$0 < 1 - e^{\lambda_0 r_0} \leq \ell \leq 1 - e^{\lambda_0 (r_0 + \mu)},$$

depending on whether $\text{val}||a_{ij}|| \leq 0$ or > 0.

Thus, if the "money" game is in his favor, Player I can defeat even an arbitrarily rich opponent, with some probability.

COROLLARY 4. If r_0 and $R \to \infty$ in a fixed ratio, or, equivalently, if the a_{ij} all $\to 0$ in a fixed ratio, then the limit of $v(r_0)$ is either 0, r_0/R, or 1, depending on whether $\text{val}||a_{ij}|| < 0$, $= 0$, or > 0, respectively.

As we pass to the limit in this fashion, the "naive" strategy of maximizing the minimum expected money gain on each round becomes better and better. Indeed, in the balanced case ($\text{val}||a_{ij}|| = 0$) it is a local F-strategy, and in the lopsided cases ($\text{val}||a_{ij}|| <$ or > 0) one player has nothing to lose anyway, in the limit, while any strategy with positive expected gain wins for the other. These remarks may clarify the rather puzzling (and not entirely correct) conclusions of [1], [2], [11] to the effect that the "naive" strategy just mentioned is approximately optimal.

[The following is an example of a game in which this "naive" strategy is not satisfactory:

$$\begin{pmatrix} -\varepsilon & -\varepsilon \\ -10 & 9 \\ 9 & -10 \end{pmatrix}$$

<u>Example 10</u>

In fact if Player I follows it here he will always choose the first row and hence always lose (assuming $0 < \varepsilon < \frac{1}{2}$). Another case where the "naive" strategy is not satisfactory for Player I is given by Example 7 above.]

A generalization of Corollary 4 has been obtained by Scarf [12] for survival games in which r and a_{ij} are n-dimensional vectors. Under certain assumptions, which reduce to our condition val$||a_{ij}|| = 0$, he finds that the limiting value functions are generalized harmonic functions, being the zeros of certain second-order differential operators, in general nonlinear.

A different extension of the survival game model, of some interest, is obtained by changing the information pattern, disrupting in some specified way the process whereby the players learn of each other's past moves and the resulting winnings or losses (see [14]). Since the local F-strategies can be played without benefit of any information whatever, the bounds of Theorem 6 remain applicable, and we have:

COROLLARY 5. In a game of survival with restricted information flow, the value (if it exists) lies within the bounds (16). In any case, the minorant (sup-inf) and majorant (inf-sup) values exist and satisfy (16).

We note in passing that the value always exists if $Q \equiv 0$ or $Q \equiv 1$, since the payoff as a function of the pure strategies is semi-continuous, and the pure strategy spaces are compact, regardless of the information pattern (compare [13]).

So far in this section we have been proceeding on the assumption that $||a_{ij}||$ is zero-free. We now indicate without proof the modifications required if this assumption is dropped. The parallel numbering will assist comparison.

LEMMA 12^0. If $\max_i \min_j a_{ij} < 0 < \min_j \max_i a_{ij}$ then the solutions of

$$\text{val } ||f(\lambda, a_{ij})|| = 0 \tag{15^0}$$

constitute a finite, closed interval $[\lambda', \lambda'']$. Moreover λ' and λ'' both have signs opposite to val$||a_{ij}||$, and we have $\lambda' \leq 0 \leq \lambda''$ if and only if val$||a_{ij}|| = 0$.

Write $F'(x)$ for $f(\lambda', x)$ and $F''(x)$ for $f(\lambda'', x)$.

LEMMA 13^0. Both F' and F'' are solutions of (3).

COROLLARY. The local F'- and F''-strategies are precisely those mixed strategies that use only probability distributions that are optimal in the matrix games $||F'(a_{ij})||$ and $||F''(a_{ij})||$ respectively.

THEOREM 6^0. If

$$\max_i \min_j a_{ij} < 0 < \min_j \max_i a_{ij}$$

then the extreme solutions v_0 and v_1 of (3), (4) are approximated by $F''(r)/F''(R)$ and $F'(r)/F'(R)$ respectively, with precise bounds of the form (16). In the $Q \equiv 0$ game, Player II can enforce the upper bound to v_0 by playing a local F''-strategy, and Player I can enforce to within any $\varepsilon > 0$ of the lower bound by choosing $\delta > 0$ small enough and playing optimal strategies of $||f(\lambda'' + \delta, a_{ij})||$ on each round. A similar statement holds for the $Q \equiv 1$ game and its value function v_1. For general Q the value (if it exists) lies between $F''(r_0)/F''(R+\nu)$ and $F'(r+\mu)/F'(R+\mu)$.

Again we remark that the cases $max\,min\,a_{ij} > 0$ and $min\,max\,a_{ij} < 0$ are trivial. A guide to what happens when one or both is equal to zero is provided by Lemma 3, in Section 2.

The five corollaries are unchanged or are modified in the obvious way, using the last part of Lemma 14^0 and noting that statements must be made in terms of v_0 and v_1, with the value of the game in general (if it exists) lying in between. Corollary 2^0 can be extended slightly (with the aid of Theorem 2) to yield the following result:

THEOREM 7. If the a_{ij} are all ± 1 or 0, and if $\lambda' = \lambda''$, then $v_0 = v_1$ and the value of the survival game exists and is independent of Q.

It is natural to ask whether $\lambda' = \lambda''$ implies $v_0 = v_1$ under more general conditions. In view of Example 11, discussed below, the answer seems to be in the negative. However the converse implication is valid almost always. In fact, if $\lambda' < \lambda''$ then the inequality:

$$\frac{F''(r+\mu)}{F''(R+\mu)} < \frac{F'(r)}{F'(R+\nu)}$$

holds at $r = R/2$ if R is sufficiently large. This implies $v_0(R/2) < v_1(R/2)$ by Theorem 6^0. This proves:

THEOREM 8. If $||a_{ij}||$ is such that the functions v_0 and v_1 are identical for large values of R, then $\lambda' = \lambda''$.

(To see that the condition on R is needed, go back to Example 2 in Section 1 and put $R = 1$. Then we have $v_0 = v_1 \equiv 1/2$ in $(0, R)$; however $\lambda' < 0 < \lambda''$.)

$$\begin{pmatrix} 0 & 0 & 0 & -2 & 1 & 1 \\ 0 & 0 & 0 & 1 & -2 & 1 \\ 0 & 0 & 0 & 1 & 1 & -2 \\ 2 & -1 & -1 & 0 & 0 & 0 \\ -1 & 2 & -1 & 0 & 0 & 0 \\ -1 & -1 & 2 & 0 & 0 & 0 \end{pmatrix}$$

Example 11

[In our final example, it is easily checked that $\text{val}||a_{ij}|| = 0$ and that $\lambda' = \lambda'' = 0$. To show that v_0 and v_1 are different let u' and u'' be the value functions for the survival games which correspond to the submatrices:

$$\begin{pmatrix} 2 & -1 & -1 \\ -1 & 2 & -1 \\ -1 & -1 & 2 \end{pmatrix} \text{ and } \begin{pmatrix} -2 & 1 & 1 \\ 1 & -2 & 1 \\ 1 & 1 & -2 \end{pmatrix}$$

respectively. Clearly $v_1 \geq max[u', u'']$ and $v_0 \leq min[u', u'']$. However, a simple calculation shows that u' and u'' are distinct for $R > 1$; hence v_0 and v_1 are also distinct.]

BIBLIOGRAPHY

1. BELLMAN, R., "On multi-stage games with imprecise payoff," RAND Corporation Research Memorandum RM-1337, September 1954.

2. BELLMAN, R., "Decision-making in the face of uncertainty - II," Naval Research Logistics Quarterly 1 (1954), pp. 327-332.

3. BELLMAN, R., and LASALLE, J., "On non-zero-sum games and stochastic processes," RAND Corporation Research Memorandum RM-212, August 1949.

4. BLACKWELL, D., "On multi-component attrition games," Naval Research Logistics Quarterly 1, (1954), pp. 210-216.

5. DOOB, J. L., Stochastic Processes, Wiley, 1953.

6. EVERETT, H., "Recursive games," this Study.

7. FELLER, W., Probability Theory and its Applications, Wiley, 1950.

8. GALE, D., and STEWART, F. M., "Infinite games with perfect information," Annals of Mathematics Study No. 28 (Princeton, 1953), pp. 245-266.

9. GLICKSBERG, I., "Minimax theorem for upper and lower semi-continuous payoffs," RAND Corporation Research Memorandum RM-478, October 1950.

10. HAUSNER, M., "Games of survival and optimal strategies in games of survival," RAND Corporation Research Memoranda RM-776 and RM-777, February 1952.

11. PEISAKOFF, M. P., "More on games of survival," RAND Corporation Research Memorandum RM-884, June 1952; reproduced in R. Bellman: "The theory of dynamic programming," RAND Corporation Report R-245 (1953), pp. 87-96.

12. SCARF, H. E., "On differential games with survival payoff," this Study.

13. SCARF, H. E., and SHAPLEY, L. S., "Games with information lag," RAND Corporation Research Memorandum RM-1320, August 1954.

14. SCARF, H. E., and SHAPLEY, L. S., "Games with partial information," this Study.

15. WALD, A., "On cumulative sums of random variables," Annals of Mathematical Statistics 15 (1944), pp. 283-296.

J. Milnor
L.S. Shapley

Princeton University
The RAND Corporation
and
The California Institute of Technology

THEORY AND DECISION LIBRARY

SERIES C: GAME THEORY, MATHEMATICAL PROGRAMMING AND OPERATIONS RESEARCH

Editor: S.H. Tijs, *University of Nijmegen, The Netherlands*

1. B.R. Munier and M.F. Shakun (eds.): *Compromise, Negotiation and Group Decision.* 1988 ISBN 90-277-2625-6
2. R. Selten: *Models of Strategic Rationality.* 1988 ISBN 90-277-2663-9
3. T. Driessen: *Cooperative Games, Solutions and Applications.* 1988 ISBN 90-277-2729-5
4. P.P. Wakker: *Additive Representations of Preferences.* A New Foundation of Decision Analysis. 1989 ISBN 0-7923-0050-5
5. A. Rapoport: *Experimental Studies of Interactive Decisions.* 1990 ISBN 0-7923-0685-6
6. K.G. Ramamurthy: *Coherent Structures and Simple Games.* 1990 ISBN 0-7923-0869-7
7. T.E.S. Raghavan, T.S. Ferguson, T. Parthasarathy and O.J. Vrieze (eds.): *Stochastic Games and Related Topics.* In Honor of Professor L.S. Shapley. 1991 ISBN 0-7923-1016-0

KLUWER ACADEMIC PUBLISHERS – DORDRECHT / BOSTON / LONDON

GPSR Compliance

The European Union's (EU) General Product Safety Regulation (GPSR) is a set of rules that requires consumer products to be safe and our obligations to ensure this.

If you have any concerns about our products, you can contact us on ProductSafety@springernature.com

In case Publisher is established outside the EU, the EU authorized representative is:

Springer Nature Customer Service Center GmbH
Europaplatz 3
69115 Heidelberg, Germany

Batch number: 10761599

Printed by Printforce, the Netherlands